AF572275

BUILDING CONSTRUCTION HANDBOOK

BUILDING CONSTRUCTION HANDBOOK

By
Sanjeev Mathur

2012

SBS Publishers & Distributors Pvt. Ltd.
New Delhi

ISBN 13 : 9789380090573

First Published in 2012

Published by:

SBS PUBLISHERS & DISTRIBUTORS PVT. LTD.
2/9, Ground Floor, Ansari Road, Darya Ganj,
New Delhi - 110002,
INDIA
Tel: 0091.11.23289119 / 41563911 / 32945311
Email: mail@sbspublishers.com
www.sbspublishers.com

Printed in India by Chaman Enterprises, New Delhi.

Preface

In the fields of architecture and civil engineering, construction is a process that consists of the building or assembling of infrastructure. Far from being a single activity, large scale construction is a feat of human multitasking. For the successful execution of a project, effective planning is essential. Building construction is the process of adding structure to real property. The vast majority of building construction projects are small renovations, such as addition of a room, or renovation of a bathroom. Often, the owner of the property acts as laborer, paymaster, and design team for the entire project. However, all building construction projects include some elements in common - design, financial, estimating and legal considerations. The modern trend in design is toward integration of previously separated specialties, especially among large firms. In the past, architects, interior designers, engineers, developers, construction managers, and general contractors were more likely to be entirely separate companies, even in the larger firms. Many projects of varying sizes reach undesirable end results, such as structural collapse, cost overruns, and/or litigation reason; those with experience in the field make detailed plans and maintain careful oversight during the project to ensure a positive outcome. Commercial building construction is procured privately or publicly utilizing various delivery methodologies, including cost estimating, hard bid, negotiated price, traditional, management contracting, construction management-at-risk, design & build and design-build bridging.

Author

Content

1

Principle and Planning

INTRODUCTION

Like the five blind men encountering different parts of an elephant, each of the numerous participants in the process of planning, designing, financing, constructing and operating physical facilities has a different perspective on project management for construction. Specialized knowledge can be very beneficial, particularly in large and complicated projects, since experts in various specialties can provide valuable services.

However, it is advantageous to understand how the different parts of the process fit together. Waste, excessive cost and delays can result from poor coordination and communication among specialists. It is particularly in the interest of owners to insure that such problems do not occur. And it behooves all participants in the process to heed the interests of owners because, in the end, it is the owners who provide the resources and call the shots.

By adopting the viewpoint of the owners, we can focus our attention on the complete process of *project management* for constructed facilities rather than the historical roles of various specialists such as planners, architects, engineering designers, constructors, fabricators, material suppliers, financial analysts and others. To be sure, each specialty has made important advances in developing new techniques and tools for efficient implementation of construction projects. However, it is through the understanding of the entire process of project management that these specialists can respond more effectively to the owner's desires for their services, in marketing their specialties, and in improving the productivity and quality of their work.

The introduction of innovative and more effective project management for construction is not an academic exercise. As reported by the "Construction Industry Cost Effectiveness Project" of the Business Roundtable: By common consensus and every available measure, the United States no longer gets it's money's worth in construction, the nation's largest industry... The creeping erosion of construction efficiency and productivity is bad news for the entire U.S. economy. Construction is a particularly seminal industry. The price of every factory, office building, hotel or power plant that is built affects the price that must be charged for the goods or services produced in it or by it. And that effect generally persists for decades... Too much of the industry remains tethered to the past, partly by inertia and partly by historic divisions.

Improvement of project management not only can aid the construction industry, but may also be the engine for the national and world economy. However, if we are to make meaningful improvements, we must first understand the construction industry, its operating environment and the institutional constraints affecting its activities as well as the nature of project management.

PROJECT LIFE CYCLE

The acquisition of a constructed facility usually represents a major capital investment, whether its owner happens to be an individual, a private corporation or a public agency. Since the commitment of resources for such an investment is motivated by market demands or perceived needs, the facility is expected to satisfy certain objectives within the constraints specified by the owner and relevant regulations. With the exception of the speculative housing market, where the residential units may be sold as built by the real estate developer, most constructed facilities are custom made in consultation with the owners. A real estate developer may be regarded as the sponsor of building projects, as much as a government agency may be the sponsor of a public project and turns it over to another government unit upon its completion.

From the viewpoint of project management, the terms "owner" and "sponsor" are synonymous because both have the ultimate authority to make all important decisions. Since an owner is essentially acquiring a facility on a promise in some form of agreement, it will be wise for any owner to have a clear understanding of the acquisition process in order to maintain firm control of the quality, timeliness and cost of the completed facility.

From the perspective of an owner, the project life cycle for a constructed facility may be illustrated schematically in Figure 1.1. Essentially, a project is conceived to meet market demands or needs in a timely fashion. Various possibilities may be considered in the conceptual planning stage, and the technological and economic feasibility of each alternative will be assessed and compared in order to select the best possible project.

The financing schemes for the proposed alternatives must also be examined, and the project will be programmed with respect to the timing for its completion and for available cash flows. After the scope of the project is clearly defined, detailed engineering design will provide the blueprint for construction, and the definitive cost estimate will serve as the baseline for cost control.

In the procurement and construction stage, the delivery of materials and the erection of the project on site must be carefully planned and controlled. After the construction is completed, there is usually a brief period of start-up or shake-down of the constructed facility when it is first occupied. Finally, the management of the facility is turned over to the owner for full occupancy until the facility lives out its useful life and is designated for demolition or conversion.

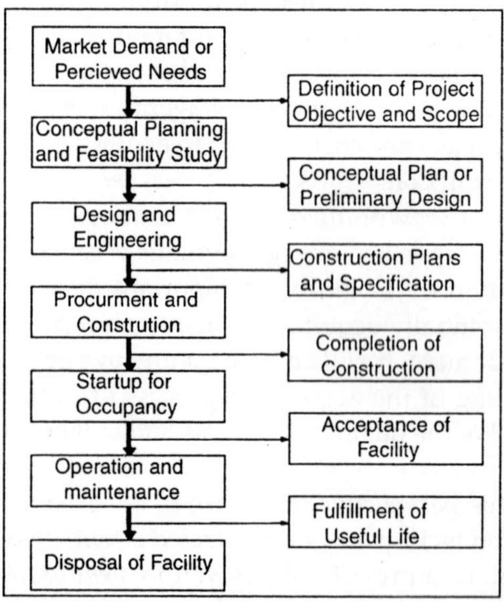

Fig. 1.1 The Project Life Cycle of a Constructed Facility

Of course, the stages of development in Figure 1.2 may not be strictly sequential. Some of the stages require iteration, and others may be carried out in parallel or with overlapping time frames, depending on the nature, size and urgency of the project. Furthermore, an owner may have in-house capacities to handle the work in every stage of the entire process, or it may seek professional advice and services for the work in all stages. Understandably, most owners choose to handle some of the work in-house and to contract outside professional services for other components of the work as needed.

By examining the project life cycle from an owner's perspective we can focus on the proper roles of various activities and participants in all stages regardless of the contractual arrangements for different types of work. In the United States, for example, the U.S. Army Corps of Engineers has in-house capabilities to deal with planning, budgeting, design, construction and operation of waterway and flood control structures. Other public agencies, such as state transportation departments, are also deeply involved in all phases of a construction project.

In the private sector, many large firms such as DuPont, Exxon, and IBM are adequately staffed to carry out most activities for plant expansion. All these owners, both public and private, use outside agents to a greater or lesser degree when it becomes more advantageous to do so.

The project life cycle may be viewed as a process through which a project is implemented from cradle to grave. This process is often very complex; however, it can be decomposed into several stages as indicated by the general outline in Figure 1.1.

The solutions at various stages are then integrated to obtain the final outcome. Although each stage requires different expertise, it usually includes both technical and managerial activities in the *knowledge domain* of the specialist. The owner may choose to decompose the entire process into more or less stages based on the size and nature of the project, and thus obtain the most efficient result in implementation.

Very often, the owner retains direct control of work in the planning and programming stages, but increasingly outside planners and financial experts are used as consultants because of the complexities of projects. Since operation and maintenance of a facility will go on long after the completion and acceptance of a project, it is usually treated as a separate problem except in the consideration of the life cycle cost of a facility. All stages from conceptual planning and feasibility studies to the acceptance of a facility for occupancy may be broadly lumped together and referred to as the Design/Construct process, while the procurement and construction alone are traditionally regarded as the province of the construction industry. Owners must recognize that there is no single best approach in organizing project management throughout a project's life cycle. All organizational approaches have advantages and disadvantages, depending on the knowledge of the owner in construction management as well as the type, size and location of the project. It is important for the owner to be aware of the approach which is most appropriate and beneficial for a particular project. In making choices, owners should be concerned with the life cycle costs of constructed facilities rather than simply the initial construction costs. Saving small amounts of money during construction may not be worthwhile if the result is much larger operating costs or not meeting the functional requirements for the new facility satisfactorily. Thus, owners must be very concerned with the quality of the finished product as well as the cost of construction itself. Since facility operation and maintenance is a part of the project life cycle, the owners' expectation to satisfy investment objectives during the project life cycle will require consideration of the cost of operation and maintenance. Therefore, the facility's operating management should also be considered as early as possible, just as the construction process should be kept in mind at the early stages of planning and programming.

MAJOR TYPES OF CONSTRUCTION

Since most owners are generally interested in acquiring only a specific type of constructed facility, they should be aware of the common industrial practices for the type of construction pertinent to them. Likewise, the *construction industry* is a conglomeration of quite diverse segments and products. Some owners may procure a constructed facility only once in a long while and tend to look for short term advantages. However, many owners require periodic acquisition of new facilities and/or rehabilitation of existing facilities. It is to their advantage to keep the construction industry healthy and productive. Collectively, the owners have more power to influence the construction industry than they realise because, by their individual actions, they can provide incentives or disincentives for innovation, efficiency and quality in construction. It is to the interest of all parties that the owners take an active interest in the construction and exercise beneficial influence on the performance of the industry.

In planning for various types of construction, the methods of procuring professional services, awarding construction contracts, and financing the constructed facility can be quite different. For the purpose of discussion, the broad spectrum of constructed facilities may be classified into four major categories, each with its own characteristics.

RESIDENTIAL HOUSING CONSTRUCTION

Residential housing construction includes single-family houses, multi-family dwellings, and high-rise apartments. During the development and construction of such projects, the developers or sponsors who are familiar with the construction industry usually serve as surrogate owners and take charge, making necessary contractual agreements for design and construction, and arranging the financing and sale of the completed structures. Residential housing designs are usually performed by architects and engineers, and the construction executed by builders who hire subcontractors for the structural, mechanical, electrical and other specialty work. An exception to this pattern is for single-family houses which may be designed by the builders as well. The residential housing market is heavily affected by general economic conditions, tax laws, and the monetary and fiscal policies of the government. Often, a slight increase in total demand will cause a substantial investment in construction, since many housing projects can be started at different locations by different individuals and developers at the same time. Because of the relative ease of entry, at least at the lower end of the market, many new builders are attracted to the residential housing construction. Hence, this market is highly competitive, with potentially high risks as well as high rewards.

Fig. 1.2 Residential Housing Construction

INSTITUTIONAL AND COMMERCIAL BUILDING CONSTRUCTION

Institutional and commercial building construction encompasses a great variety of project types and sizes, such as schools and universities, medical clinics and hospitals, recreational facilities and sports stadiums, retail chain stores and large shopping centers, warehouses and light manufacturing plants, and skyscrapers for offices and hotels. The owners of such buildings may or may not be familiar with construction industry practices, but they usually are able to select competent professional consultants and arrange the financing of the constructed facilities themselves. Specialty architects and engineers are often engaged for designing a specific type of building, while the builders or general contractors undertaking such projects may also be specialized in only that type of building.

Fig. 1.3 Construction of the PPG Building in Pittsburgh

Because of the higher costs and greater sophistication of institutional and commercial buildings in comparison with residential housing, this market segment is shared by fewer competitors. Since the construction of some of these buildings is a long process which once started will take some time to proceed until completion, the demand is less sensitive to general economic conditions than that for speculative housing. Consequently, the owners may confront an *oligopoly* of general contractors who compete in the same market. In an oligopoly situation, only a limited number of competitors exist, and a firm's price for services may be based in part on its competitive strategies in the local market.

SPECIALIZED INDUSTRIAL CONSTRUCTION

Specialized industrial construction usually involves very large scale projects with a high degree of technological complexity, such as oil refineries, steel mills, chemical processing plants and coal-fired or nuclear power plants. The owners usually are deeply involved in the development of a project, and prefer to work with designers-builders such that the total time for the completion of the project can be shortened. They also want to pick a team of designers and builders with whom the owner has developed good working relations over the years. Although the initiation of such projects is also affected by the state of the economy, long range demand forecasting is the most important factor since such projects are capital intensive and require considerable amount of planning and construction time. Governmental regulation such as the rulings of the Environmental Protection Agency and the Nuclear Regulatory Commission in the United States can also profoundly influence decisions on these projects.

Fig. 1.4 Construction of a Benzene Plant in Lima

INFRASTRUCTURE AND HEAVY CONSTRUCTION

Infrastructure and heavy construction includes projects such as highways, mass transit systems, tunnels, bridges, pipelines, drainage systems and sewage treatment plants. Most of these projects are publicly owned and therefore financed either through bonds or taxes. This category of construction is characterized by a high degree of mechanization, which has gradually replaced some labor intensive operations. The engineers and builders engaged in infrastructure construction are usually highly specialized since each segment of the market requires different types of skills.

However, demands for different segments of infrastructure and heavy construction may shift with saturation in some segments. For example, as the available highway construction projects are declining, some heavy construction contractors quickly move their work force and equipment into the field of mining where jobs are available.

Fig. 1.5 Construction of the Dame Point Bridge in Jacksonville

SELECTION OF PROFESSIONAL SERVICES

When an owner decides to seek professional services for the design and construction of a facility, he is confronted with a broad variety of choices. The type of services selected depends to a large degree on the type of construction and the experience of the owner in dealing with various professionals in the previous projects undertaken by the firm. Generally, several common types of professional services may be engaged either separately or in some combination by the owners.

FINANCIAL PLANNING CONSULTANTS

At the early stage of strategic planning for a capital project, an owner often seeks the services of financial planning consultants such as certified public accounting (CPA) firms to evaluate the economic and financial feasibility of the constructed facility, particularly with respect to various provisions of federal, state and local tax laws which may affect the investment decision. Investment banks may also be consulted on various options for financing the facility in order to analyse their long-term effects on the financial health of the owner organization.

ARCHITECTURAL AND ENGINEERING FIRMS

Traditionally, the owner engages an architectural and engineering (A/E) firm or consortium as technical consultant in developing a preliminary design. After the engineering design and financing arrangements for the project are completed, the owner will enter into a construction contract with a general contractor either through competitive bidding or negotiation. The general contractor will act as a constructor and/or a coordinator of a large number of subcontractors who perform various specialties for the completion of the project. The A/E firm completes the design and may also provide on site quality inspection during construction. Thus, the A/E firm acts as the prime professional on behalf of the owner and supervises the construction to insure satisfactory results. This practice is most common in building construction.

In the past two decades, this traditional approach has become less popular for a number of reasons, particularly for large scale projects. The A/E firms, which are engaged by the owner as the prime professionals for design and inspection, have become more isolated from the construction process. This has occurred because of pressures to reduce fees to A/E firms, the threat of litigation regarding construction defects, and lack of knowledge of new construction techniques on the part of architect and engineering professionals. Instead of preparing a construction plan along with the design, many A/E firms are no longer responsible for the details of construction nor do they provide periodic field inspection in many cases. As a matter of fact, such firms will place a prominent disclaimer of responsibilities on any shop drawings they may check, and they will often regard their representatives in the field as observers instead of inspectors. Thus, the A/E firm and the general contractor on a project often become antagonists who are looking after their own competing interests. As a result, even the constructability of some engineering designs may become an issue of contention. To carry this protective attitude to the extreme, the specifications prepared by an A/E firm for the general contractor often protects the interest of the A/E firm at the expense of the interests of the owner and the contractor.

In order to reduce the cost of construction, some owners introduce *value engineering,* which seeks to reduce the cost of construction by soliciting a second design that might cost less than the original design produced by the A/E firm. In practice, the second design is submitted by the contractor after receiving a construction contract at a stipulated sum, and the saving in cost resulting from the redesign is shared by the contractor and the owner. The contractor is able to absorb the cost of redesign from the profit in construction or to reduce the construction cost as a result of the re-design. If the owner had been willing to pay a higher fee to the A/E firm or to better direct the design process, the A/E firm might have produced an improved design which would cost less in the first place. Regardless of the merit of value engineering, this practice has undermined the role of the A/E firm as the prime professional acting on behalf of the owner to supervise the contractor.

DESIGN/CONSTRUCT FIRMS

A common trend in industrial construction, particularly for large projects, is to engage the services of a design/construct firm. By integrating design and construction management in a single organization, many of the conflicts between designers and constructors might be avoided. In particular, designs will be closely scrutinized for their constructibility. However, an owner engaging a design/construct firm must insure that the quality of the constructed facility is not sacrificed by the desire to reduce the time or the cost for completing the project. Also, it is difficult to make use of competitive bidding in this type of design/construct process. As a result, owners must be relatively sophisticated in negotiating realistic and cost-effective construction contracts.

One of the most obvious advantages of the integrated design/construct process is the use of *phased construction* for a large project. In this process, the project is divided up into several phases, each of which can be designed and constructed in a staggered manner. After the completion of the design of the first phase, construction can begin without waiting for the completion of the design of the second phase, etc. If proper coordination is exercised. the total project duration can be greatly reduced. Another advantage is to exploit the possibility of using the *turnkey* approach whereby an owner can delegate all responsibility to the design/construct firm which will deliver to the owner a completed facility that meets the performance specifications at the specified price.

PROFESSIONAL CONSTRUCTION MANAGERS

In recent years, a new breed of construction managers (CM) offers professional services from the inception to the completion of a construction project. These construction managers mostly come from the ranks of A/E firms or general contractors who may or may not retain dual roles in the service of the owners. In any case, the owner can rely on the service of a single prime professional to manage the entire process of a construction project. However, like the A/E firms of several decades ago, the construction managers are appreciated by some owners but not by others. Before long, some owners find that the construction managers too may try to protect their own interest instead of that of the owners when the stakes are high.

It should be obvious to all involved in the construction process that the party which is required to take higher risk demands larger rewards. If an owner wants to engage an A/E firm on the basis of low fees instead of established qualifications, it often gets what it deserves; or if the owner wants the general contractor to bear the cost of uncertainties in construction such as foundation conditions, the contract price will be higher even if competitive bidding is used in reaching a contractual agreement. Without mutual respect and trust, an owner cannot expect that construction managers can produce better results than other professionals. Hence, an owner must understand its own responsibility and the risk it wishes to assign to itself and to other participants in the process.

OPERATION AND MAINTENANCE MANAGERS

Although many owners keep a permanent staff for the operation and maintenance of constructed facilities, others may prefer to contract such tasks to professional managers. Understandably, it is common to find in-house staff for operation and maintenance in specialized industrial plants and infrastructure facilities, and the use of outside managers under contracts for the operation and maintenance of rental properties such as apartments and office buildings. However, there are exceptions to these common practices. For example, maintenance of public roadways can be contracted to private firms.

In any case, managers can provide a spectrum of operation and maintenance services for a specified time period in accordance to the terms of contractual agreements. Thus, the owners can be spared the provision of in-house expertise to operate and maintain the facilities.

FACILITIES MANAGEMENT

As a logical extension for obtaining the best services throughout the project life cycle of a constructed facility, some owners and developers are receptive to adding strategic planning at the beginning and facility maintenance as a follow-up to reduce space-related costs in their real estate holdings. Consequently, some architectural/engineering firms and construction management firms with computer-based expertise, together with interior design firms, are offering such front-end and follow-up services in addition to the more traditional services in design and construction.

Facilities management is the discipline of planning, designing, constructing and managing space — in every type of structure from office buildings to process plants. It involves developing corporate facilities policy, long-range forecasts, real estate, space inventories, projects (through design, construction and renovation), building operation and maintenance plans and furniture and equipment inventories.

A common denominator of all firms entering into these new services is that they all have strong computer capabilities and heavy computer investments. In addition to the use of computers for aiding design and monitoring construction, the service includes the compilation of a computer record of building plans that can be turned over at the end of construction to the facilities management group of the owner. A computer data base of facilities information makes it possible for planners in the owner's organization to obtain overview information for long range space forecasts, while the line managers can use as-built information such as lease/tenant records, utility costs, etc. for day-to-day operations.

CONSTRUCTION CONTRACTORS

Builders who supervise the execution of construction projects are traditionally referred to as *contractors*, or more appropriately called *constructors*. The *general contractor* coordinates various tasks for a project while the *specialty contractors* such as mechanical or electrical contractors perform the work in their specialties. Material and equipment suppliers often act as *installation contractors*; they play a significant role in a construction project since the conditions of delivery of materials and equipment affect the quality, cost, and timely completion of the project. It is essential to understand the operation of these contractors in order to deal with them effectively.

GENERAL CONTRACTORS

The function of a general contractor is to coordinate all tasks in a construction project. Unless the owner performs this function or engages a professional construction manager to do so, a good general contractor who has worked with a team of superintendents, specialty contractors or subcontractors together for a number of projects in the past can be most effective in inspiring loyalty and cooperation. The general contractor is also knowledgeable about the labor force employed in construction. The labor force may or may not be unionized depending on the size and location of the projects. In some projects, no member of the work force belongs to a labor union; in other cases, both union and non-union craftsmen work together in what is called an open shop, or all craftsmen must be affiliated with labor unions in a closed shop. Since labor unions provide hiring halls staffed with skilled journeyman who have gone through apprentice programmes for the projects as well as serving as collective bargain units, an experienced general contractor will make good use of the benefits and avoid the pitfalls in dealing with organized labor.

SPECIALTY CONTRACTORS

Specialty contractors include mechanical, electrical, foundation, excavation, and demolition contractors among others. They usually serve as subcontractors to the general contractor of a project. In some cases, legal statutes may require an owner to deal with various specialty contractors directly. In the State of New York, for example, specialty contractors, such as mechanical and electrical contractors, are not subjected to the supervision of the general contractor of a construction project and must be given separate prime contracts on public works. With the exception of such special cases, an owner will hold the general contractor responsible for negotiating and fulfilling the contractual agreements with the subcontractors.

MATERIAL AND EQUIPMENT SUPPLIERS

Major material suppliers include specialty contractors in structural steel fabrication and erection, sheet metal, ready mixed concrete delivery, reinforcing steel bar detailers, roofing, glazing etc. Major equipment suppliers for industrial construction include manufacturers of generators, boilers and piping and other equipment. Many suppliers handle on-site installation to insure that the requirements and contractual specifications are met. As more and larger structural units are prefabricated off-site, the distribution between specialty contractors and material suppliers becomes even less obvious.

FINANCING OF CONSTRUCTED FACILITIES

A major construction project requires an enormous amount of capital that is often supplied by lenders who want to be assured that the project will offer a fair return on the investment.

The direct costs associated with a major construction project may be broadly classified into two categories:

- The construction expenses paid to the general contractor for erecting the facility on site and
- The expenses for land acquisition, legal fees, architect/ engineer fees, construction management fees, interest on construction loans and the opportunity cost of carrying empty space in the facility until it is fully occupied.

The direct construction costs in the first category represent approximately 60 to 80 percent of the total costs in most construction projects. Since the costs of construction are ultimately borne by the owner, careful financial planning for the facility must be made prior to construction.

CONSTRUCTION FINANCING

Construction loans to contractors are usually provided by banks or savings and loan associations for construction financing. Upon the completion of the facility, construction loans will be terminated and the post-construction facility financing will be arranged by the owner.

Construction loans provided for different types of construction vary. In the case of residential housing, construction loans and long-term mortgages can be obtained from savings and loans associations or commercial banks. For institutional and commercial buildings, construction loans are usually obtained from commercial banks. Since the value of specialized industrial buildings as collateral for loans is limited, construction loans in this domain are rare, and construction financing can be done from the pool of general corporate funds.

For infrastructure construction owned by government, the property cannot be used as security for a private loan, but there are many possible ways to finance the construction, such as general appropriation from taxation or special bonds issued for the project.

Traditionally, banks serve as construction lenders in a three-party agreement among the contractor, the owner and the bank. The stipulated loan will be paid to the contractor on an agreed schedule upon the verification of completion of various portions of the project.

Generally, a payment request together with a standard progress report will be submitted each month by the contractor to the owner which in turn submits a draw request to the bank. Provided that the work to date has been performed satisfactorily, the disbursement is made on that basis during the construction period. Under such circumstances, the bank has been primarily concerned with the completion of the facility on time and within the budget. The economic life of the facility after its completion is not a concern because of the transfer of risk to the owner or an institutional lender.

FACILITY FINANCING

Many private corporations maintain a pool of general funds resulting from retained earnings and long-term borrowing on the strength of corporate assets, which can be used for facility financing. Similarly, for public agencies, the long-term funding may be obtained from the commitment of general tax revenues from the federal, state and/or local governments. Both private corporations and public agencies may issue special bonds for the constructed facilities which may obtain lower interest rates than other forms of borrowing.

Short-term borrowing may also be used for bridging the gaps in long-term financing. Some corporate bonds are convertible to stocks under circumstances specified in the bond agreement. For public facilities, the assessment of user fees to repay the bond funds merits consideration for certain types of facilities such as toll roads and sewage treatment plants. The use of mortgages is primarily confined to rental properties such as apartments and office buildings.

Because of the sudden surge of interest rates in the late 1970's, many financial institutions offer, in addition to the traditional fixed rate long-term mortgage commitments, other arrangements such as a combination of debt and a percentage of ownership in exchange for a long-term mortgage or the use of adjustable rate mortgages. In some cases, the construction loan may be granted on an open-ended basis without a long-term financing commitment.

For example, the plan might be issued for the construction period with an option to extend it for a period of up to three years in order to give the owner more time to seek alternative long-term financing on the completed facility. The bank will be drawn into situations involving financial risk if it chooses to be a lender without long-term guarantees.

For international projects, the currency used for financing agreements becomes important. If financial agreements are written in terms of local currencies, then fluctuations in the currency exchange rate can significantly affect the cost and ultimately profit of a project. In some cases, payments might also be made in particular commodities such as petroleum or the output from the facility itself. Again, these arrangements result in greater uncertainty in the financing scheme because the price of these commodities may vary.

LEGAL AND REGULATORY REQUIREMENTS

The owners of facilities naturally want legal protection for all the activities involved in the construction. It is equally obvious that they should seek competent legal advice. However, there are certain principles that should be recognized by owners in order to avoid unnecessary pitfalls.

LEGAL RESPONSIBILITIES

Activities in construction often involve risks, both physical and financial. An owner generally tries to shift the risks to other parties to the degree possible when entering into contractual agreements with them. However, such action is not without cost or risk. For example, a contractor who is assigned the risks may either ask for a higher contract price to compensate for the higher risks, or end up in non-performance or bankruptcy as an act of desperation. Such consequences can be avoided if the owner is reasonable in risk allocation. When risks are allocated to different parties, the owner must understand the implications and spell them out clearly. Sometimes there are statutory limitations on the allocation of liabilities among various groups, such as prohibition against the allocation of negligence in design to the contractor. An owner must realise its superior power in bargaining and hence the responsibilities associated with this power in making contractual agreements.

MITIGATION OF CONFLICTS

It is important for the owner to use legal counselors as advisors to mitigate conflicts before they happen rather than to wield conflicts as weapons against other parties. There are enough problems in design and construction due to uncertainty rather than bad intentions.

The owner should recognize the more enlightened approaches for mitigating conflicts, such as using owner-controlled *wrap-up* insurance which will provide protection for all parties involved in the construction process for unforeseen risks, or using arbitration, mediation and other extra-judicial solutions for disputes among various parties. However, these compromise solutions are not without pitfalls and should be adopted only on the merit of individual cases.

GOVERNMENT REGULATION

To protect public safety and welfare, legislatures and various government agencies periodically issue regulations which influence the construction process, the operation of constructed facilities, and their ultimate disposal. For example, building codes promulgated by local authorities have provided guidelines for design and construction practices for a very long time. Since the 1970's, many federal regulations that are related directly or indirectly to construction have been established in the United States. Among them are safety standards for workers issued by the Occupational Health and Safety Administration, environmental standards on pollutants and toxic wastes issued by the Environmental Protection Agency, and design and operation procedures for nuclear power plants issued by the Nuclear Regulatory Commission.

Owners must be aware of the impacts of these regulations on the costs and durations of various types of construction projects as well as possibilities of litigation due to various contentions. For example, owners acquiring sites for new construction may be strictly liable for any hazardous wastes already on the site or removed from the site under the U.S. Comprehensive Environmental Response Compensation and Liability (CERCL) Act of 1980. For large scale projects involving new technologies, the construction costs often escalate with the uncertainty associated with such restrictions.

CHANGING ENVIRONMENT OF THE CONSTRUCTION INDUSTRY

The construction industry is a conglomeration of diverse fields and participants that have been loosely lumped together as a sector of the economy. The construction industry plays a central role in national welfare, including the development of residential housing, office buildings and industrial plants, and the restoration of the nation's infrastructure and other public facilities. The importance of the construction industry lies in the function of its products which provide the foundation for industrial production, and its impacts on the national economy cannot be measured by the value of its output or the number of persons employed in its activities alone.

To be more specific, construction refers to all types of activities usually associated with the erection and repair of immobile facilities. Contract construction consists of a large number of firms that perform construction work for others, and is estimated to be approximately 85% of all construction activities. The remaining 15% of construction is performed by owners of the facilities, and is referred to as *force-account* construction.

favorable conditions and to avoid the pitfalls. Several factors are particularly noteworthy because of their significant impacts on the quality, cost and time of construction.

NEW TECHNOLOGIES

In recent years, technological innovation in design, materials and construction methods have resulted in significant changes in construction costs. Computer-aids have improved capabilities for generating quality designs as well as reducing the time required to produce alternative designs.

New materials not only have enhanced the quality of construction but also have shortened the time for shop fabrication and field erection. Construction methods have gone through various stages of mechanization and automation, including the latest development of construction robotics.

The most dramatic new technology applied to construction has been the Internet and its private, corporate Intranet versions. The Internet is widely used as a means to foster collaboration among professionals on a project, to communicate for bids and results, and to procure necessary goods and services. Real time video from specific construction sites is widely used to illustrate construction progress to interested parties.

The result has been more effective collaboration, communication and procurement. The effects of many new technologies on construction costs have been mixed because of the high development costs for new technologies. However, it is unmistakable that design professionals and construction contractors who have not adapted to changing technologies have been forced out of the mainstream of design and construction activities. Ultimately, construction quality and cost can be improved with the adoption of new technologies which are proved to be efficient from both the viewpoints of performance and economy.

LABOR PRODUCTIVITY

The term *productivity* is generally defined as a ratio of the production output volume to the input volume of resources. Since both output and input can be quantified in a number of ways, there is no single measure of productivity that is universally applicable, particularly in the construction industry where the products are often unique and there is no standard for specifying the levels for aggregation of data.

However, since labor constitutes a large part of the cost of construction, labor productivity in terms of output volume (constant dollar value or functional units) per person-hour is a useful measure.

Labor productivity measured in this way does not necessarily indicate the efficiency of labor alone but rather measures the combined effects of labor, equipment and other factors contributing to the output.

While aggregate construction industry productivity is important as a measure of national economy, owners are more concerned about the labor productivity of basic units of work produced by various crafts on site. Thus, an owner can compare the labor performance at different geographic locations, under different working conditions, and for different types and sizes of projects.

Construction costs usually run parallel to material prices and labor wages. Actually, over the years, labor productivity has increased in some traditional types of construction and thus provides a leveling or compensating effect when hourly rates for labor increase faster than other costs in construction. However, labor productivity has been stagnant or even declined in unconventional or large scale projects.

PUBLIC SCRUTINY

Under the present litigious climate in the United States, the public is increasingly vocal in the scrutiny of construction project activities. Sometimes it may result in considerable difficulty in siting new facilities as well as additional expenses during the construction process itself. Owners must be prepared to manage such crises before they get out of control.

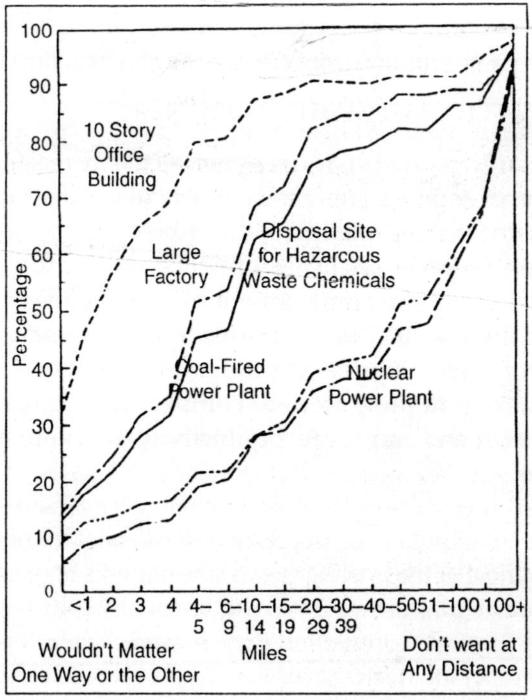

Fig. 1.8 Public Acceptance Towards New Facilities

Figure 1.8 can serve to indicate public attitudes towards the siting of new facilities. It represents the cumulative percentage of individuals who would be willing to accept a new industrial facility at various distances from their homes. For example, over fifty per cent of the people surveyed would accept a ten-story office building within five miles of their home, but only twenty-five per cent would accept a large factory or coal fired power plant at a similar distance. An even lower percentage would accept a hazardous waste disposal site or a nuclear power plant. Even at a distance of one hundred miles, a significant fraction of the public would be unwilling to accept hazardous waste facilities or nuclear power plants.

This objection to new facilities is a widespread public attitude, representing considerable skepticism about the external benefits and costs which new facilities will impose. It is this public attitude which is likely to make public scrutiny and regulation a continuing concern for the construction industry.

INTERNATIONAL COMPETITION

A final trend which deserves note is the increasing level of international competition in the construction industry. Owners are likely to find non-traditional firms bidding for construction work, particularly on large projects. Separate bids from numerous European, North American, and Asian construction firms are not unusual. In the United States, overseas firms are becoming increasingly visible and important. In this environment of heightened competition, good project management and improved productivity are more and more important.

A bidding competition for a major new offshore drilling platform illustrates the competitive environment in construction. Through most of the postwar years, the nation's biggest builders of offshore oil platforms enjoyed an unusually cozy relationship with the Big Oil Companies they served. Their top officials developed personal friendships with oil executives, entertained them at opulent hunting camps- and won contracts to build nearly every major offshore oil platform in the world....But this summer, the good-old boy network fell apart. Shell [Oil Co.] awarded the main contract for [a new] platform-taller than Chicago's Sears Tower, four times heavier than the Brooklyn Bridge-to a tiny upstart.

The winning bidder arranged overseas fabrication of the rig, kept overhead costs low, and proposed a novel assembly procedure by which construction equipment was mounted on completed sections of the platform in order to speed the completion of the entire structure. The result was lower costs than those estimated and bid by traditional firms.

Of course, U.S. firms including A/E firms, contractors and construction managers are also competing in foreign countries. Their success or failure in the international arena may also affect their capacities and vitality to provide services in the domestic U.S. market.

CONTRACTOR FINANCED PROJECTS

Increasingly, some owners look to contractors or joint ventures as a resource to design, to build and to finance a constructed facility. For example, a utility company may seek a consortium consisting of a design/construct firm and a financial investment firm to assume total liability during construction and thereby eliminate the risks of cost escalation to ratepayers, stockholders and the management. On the other hand, a local sanitation district may seek such a consortium to provide private ownership for a proposed new sewage treatment plant. In the former case, the owner may take over the completed facility and service the debt on construction through long-term financing arrangements; in the latter case, the private owner may operate the completed facility and recover its investment through user fees. The activities of joint ventures among design, construction and investment firms are sometimes referred to as *financial engineering*.

This type of joint venture has become more important in the international construction market where aggressive contractors often win contracts by offering a more attractive financing package rather than superior technology. With a deepening shadow of international debts in recent years, many developing countries are not in a position to undertake any new project without contractor-backed financing. Thus, the contractors or joint ventures in overseas projects are forced into very risky positions if they intend to stay in the competition.

LEAN CONSTRUCTION

"Lean manufacturing" had a revolutionary effect on many industries, especially automotive assembly companies.

Characteristics of this approach include:

- Improvement in quality and reduction of waste everywhere. Rather than increasing costs, reducing defects and waste proved to improve quality and reduce costs.
- Empowering workers to be responsible for satisfying customer needs. In construction, for example,

craftsman should make sure their work satisfied the design intent.

- Continuous improvement of processes involving the entire workforce.

Lean construction is intended to spread these practices within the construction industry. Of course, well managed construction projects already have many aspects of lean construction. For example, just-in-time delivery of materials is commonplace to avoid the waste of large inventory stockpiles. Green building projects attempt to re-use or recycle all construction wastes. But the systematic attention to continuous improvement and zero accidents and defects is new.

ROLE OF PROJECT MANAGERS

In the project life cycle, the most influential factors affecting the outcome of the project often reside at the early stages. At this point, decisions should be based on competent economic evaluation with due consideration for adequate financing, the prevalent social and regulatory environment, and technological considerations. Architects and engineers might specialize in planning, in construction field management, or in operation, but as project managers, they must have some familiarity with all such aspects in order to understand properly their role and be able to make competent decisions. Since the 1970's, many large-scale projects have run into serious problems of management, such as cost overruns and long schedule delays. Actually, the management of *megaprojects* or *superprojects* is not a practice peculiar to our time. Witness the construction of transcontinental railroads in the Civil War era and the construction of the Panama Canal at the turn of this century. Although the megaprojects of this generation may appear in greater frequency and present a new set of challenge, the problems are organizational rather than technical.

It is customary to think of engineering as a part of a trilogy, pure science, applied science and engineering. It needs emphasis that this trilogy is only one of a triad of trilogies into which engineering fits. This first is pure science, applied science and engineering; the second is economic theory, finance and engineering; and the third is social relations, industrial relations and engineering. Many engineering problems are as closely allied to social problems as they are to pure science.

As engineers advance professionally, they often spend as much or more time on planning, management and other economic or social problems as on the traditional engineering design and analysis problems which form the core of most educational programmes. It is upon the ability of engineers to tackle all such problems that their performance will ultimately be judged.

The greatest stumbling block to effective management in construction is the inertia and historic divisions among planners, designers and constructors. While technical competence in design and innovation remains the foundation of engineering practice, the social, economic and organizational factors that are pervasive in influencing the success and failure of construction projects must also be dealt with effectively by design and construction organizations. Of course, engineers are not expected to know every detail of management techniques, but they must be knowledgeable enough to anticipate the problems of management so that they can work harmoniously with professionals in related fields to overcome the inertia and historic divisions.

Paradoxically, engineers who are creative in engineering design are often innovative in planning and management since both types of activities involve problem solving. In fact, they can reinforce each other if both are included in the education process, provided that creativity and innovation instead of routine practice are emphasized. A project manager who is well educated in the *fundamental principles* of engineering design and management can usefully apply such principles once he or she has acquired basic understanding of a new *application area*. A project manager who has been trained by rote learning for a specific type of project may merely gain one year of experience repeated twenty times even if he or she has been in the field for twenty years. A broadly educated project manager can reasonably hope to become a leader in the profession; a narrowly trained project manager is often relegated to the role of his or her first job level permanently.

The owners have much at stake in selecting a competent project manager and in providing her or him with the authority to assume responsibility at various stages of the project regardless of the types of contractual agreements for implementing the project. Of course, the project manager must also possess the leadership quality and the ability to handle effectively intricate interpersonal relationships within an organization. The ultimate test of the education and experience of a project manager for construction lies in her or his ability to apply fundamental principles to solving problems in the new and unfamiliar situations which have become the hallmarks of the changing environment in the construction industry.

Although the number of contractors in the United States exceeds a million, over 60% of all contractor construction is performed by the top 400 contractors. The value of new construction in the United States (expressed in constant dollars) and the value of construction as a percentage of the gross national products from 1950 to 1985 are shown in Figures 1.6 and 1.7. It can be seen that construction is a significant factor in the Gross National Product although its importance has been declining in recent years. Not to be ignored is the fact that as the nation's constructed facilities become older, the total expenditure on rehabilitation and maintenance may increase relative to the value of new construction.

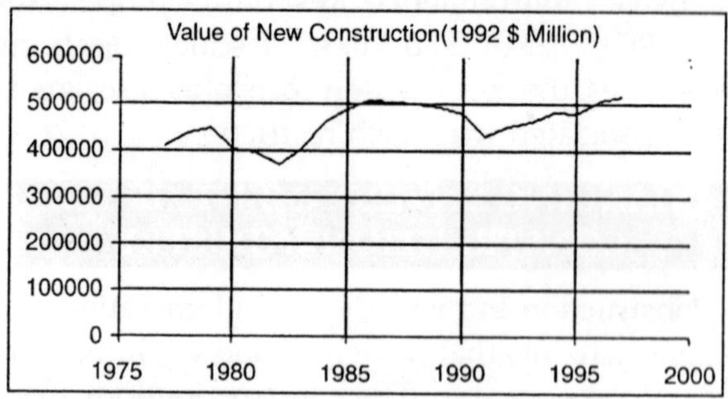

Fig. 1.6 Value of New Construction

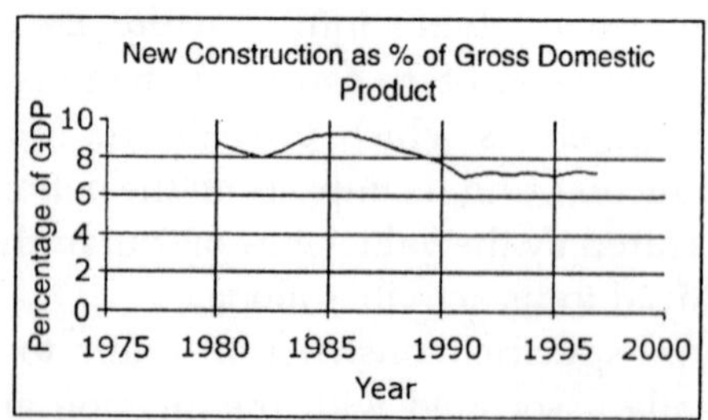

Fig. 1.7 Construction as Percentage of Gross Domestic Product

Owners who pay close attention to the peculiar characteristics of the construction industry and its changing operating environment will be able to take advantage of the

Specifically, project management in construction encompasses a set of objectives which may be accomplished by implementing a series of operations subject to resource constraints.

There are potential conflicts between the stated objectives with regard to scope, cost, time and quality, and the constraints imposed on human material and financial resources. These conflicts should be resolved at the onset of a project by making the necessary tradeoffs or creating new alternatives.

Subsequently, the functions of project management for construction generally include the following:

- Specification of project objectives and plans including delineation of scope, budgeting, scheduling, setting performance requirements, and selecting project participants.
- Maximization of efficient resource utilization through procurement of labor, materials and equipment according to the prescribed schedule and plan.
- Implementation of various operations through proper coordination and control of planning, design, estimating, contracting and construction in the entire process.
- Development of effective communications and mechanisms for resolving conflicts among the various participants.

The Project Management Institute focuses on nine distinct areas requiring project manager knowledge and attention:

- Project integration management to ensure that the various project elements are effectively coordinated.
- Project scope management to ensure that all the work required (and only the required work) is included.
- Project time management to provide an effective project schedule.
- Project cost management to identify needed resources and maintain budget control.
- Project quality management to ensure functional requirements are met.

- Project human resource management to development and effectively employ project personnel.
- Project communications management to ensure effective internal and external communications.
- Project risk management to analyse and mitigate potential risks.
- Project procurement management to obtain necessary resources from external sources.

These nine areas form the basis of the Project Management Institute's certification programme for project managers in any industry.

TRENDS IN MODERN MANAGEMENT

In recent years, major developments in management reflect the acceptance to various degrees of the following elements:

- The management process approach,
- The management science and decision support approach,
- The behavioral science approach for human resource development, and
- Sustainable competitive advantage.

These four approaches complement each other in current practice, and provide a useful groundwork for project management. The management process approach emphasizes the systematic study of management by identifying management functions in an organization and then examining each in detail. There is general agreement regarding the functions of planning, organizing and controlling. A major tenet is that by analyzing management along functional lines, a framework can be constructed into which all new management activities can be placed.

Thus, the manager's job is regarded as coordinating a process of interrelated functions, which are neither totally random nor rigidly predetermined, but are dynamic as the process evolves. Another tenet is that management principles can be derived from an intellectual analysis of management functions. By dividing the manager's job into functional components, principles based upon each function can be extracted. Hence, management functions can be organized into a hierarchical structure designed to improve operational efficiency, such as the example of the organization for a manufacturing company shown in Figure 2.2. The basic management functions are performed by all managers, regardless of enterprise, activity or hierarchical levels. Finally, the development of a management philosophy results in helping the manager to establish relationships between human and material resources. The outcome of following an established philosophy of operation helps the manager win the support of the subordinates in achieving organizational objectives.

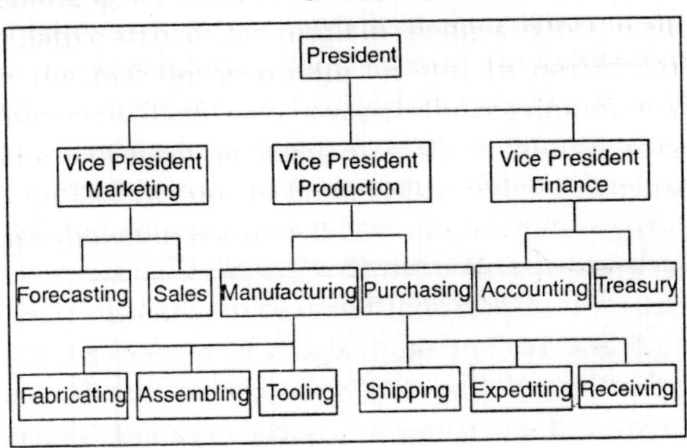

Fig. 2.2 Illustrative Hierarchical Structure of Management Functions

The management science and decision support approach contributes to the development of a body of quantitative methods designed to aid managers in making complex decisions related to operations and production. In decision support systems, emphasis is placed on providing managers with relevant information. In management science, a great deal of attention is given to defining objectives and constraints, and to constructing mathematical analysis models in solving complex problems of inventory, materials and production control, among others. A topic of major interest in management science is the maximization of profit, or in the absence of a workable model for the operation of the entire system, the sub optimization of the operations of its components. The optimization or sub optimization is often achieved by the use of operations research techniques, such as linear programming, quadratic programming, graph theory, queuing theory and Monte Carlo simulation. In addition to the increasing use of computers accompanied by the development of sophisticated mathematical models and information systems, management science and decision support systems have played an important role by looking more carefully at problem inputs and relationships and by promoting goal formulation and measurement of performance. Artificial intelligence has also begun to be applied to provide decision support systems for solving ill-structured problems in management.

The behavioral science approach for human resource development is important because management entails getting things done through the actions of people. An effective manager must understand the importance of human factors such as needs, drives, motivation, leadership, personality, behaviour, and work groups. Within this context, some place more emphasis on interpersonal behaviour which focuses on the individual and his/her motivations as a socio-psychological being; others emphasize more group behaviour in recognition of the organized enterprise as a social organism, subject to all the attitudes, habits, pressures and conflicts of the cultural environment of people.

The major contributions made by the behavioral scientists to the field of management include:

- The formulation of concepts and explanations about individual and group behaviour in the organization,
- The empirical testing of these concepts methodically in many different experimental and field settings, and
- The establishment of actual managerial policies and decisions for operation based on the conceptual and methodical frameworks.

Sustainable competitive advantage stems primarily from good management strategy. As Michael Porter of the Harvard Business School argues: Strategy is creating fit among a company's activities. The success of a strategy depends on doing many things well-not just a few-and integrating among them. If there is no fit among activites, there is no distinctive strategy and little sustainability.

In this view, successful firms must improve and align the many processes underway to their strategic vision.

Strategic positioning in this fashion requires:

- Creating a unique and valuable position.
- Making trade-offs compared to competitors
- Creating a "fit" among a company's activities.

Project managers should be aware of the strategic position of their own organization and the other organizations involved in the project. The project manager faces the difficult task of trying to align the goals and strategies of these various organizations to accomplish the project goals. For example, the owner of an industrial project may define a strategic goal as being first to market with new products. In this case, facilities development must be oriented to fast-track, rapid construction. As another example, a contracting firm may see their strategic advantage in new technologies and emphasize profit opportunities from value engineering.

STRATEGIC PLANNING AND PROJECT PROGRAMMING

The programming of capital projects is shaped by the strategic plan of an organization, which is influenced by market demands and resources constraints. The programming process associated with planning and feasibility studies sets the priorities and timing for initiating various projects to meet the overall objectives of the organizations. However, once this decision is made to initiate a project, market pressure may dictate early and timely completion of the facility.

Among various types of construction, the influence of market pressure on the timing of initiating a facility is most obvious in industrial construction. Demand for an industrial product may be short-lived, and if a company does not hit the market first, there may not be demand for its product later. With intensive competition for national and international markets, the trend of industrial construction moves towards shorter project life cycles, particularly in technology intensive industries.

In order to gain time, some owners are willing to forego thorough planning and feasibility study so as to proceed on a project with inadequate definition of the project scope. Invariably, subsequent changes in project scope will increase construction costs; however, profits derived from earlier facility operation often justify the increase in construction costs. Generally, if the owner can derive reasonable profits from the operation of a completed facility, the project is considered a success even if construction costs far exceed the estimate based on an inadequate scope definition. This attitude may be attributed in large part to the uncertainties inherent in construction projects. It is difficult to argue that profits might be even higher if construction costs could be reduced without increasing the project duration. However, some projects, notably some nuclear power plants, are clearly unsuccessful and abandoned before completion, and their demise must be attributed at least in part to inadequate planning and poor feasibility studies.

The owner or facility sponsor holds the key to influence the construction costs of a project because any decision made at the beginning stage of a project life cycle has far greater influence than those made at later stages, as shown schematically in Figure 2.3. Moreover, the design and construction decisions will influence the continuing operating costs and, in many cases, the revenues over the facility lifetime. Therefore, an owner should obtain the expertise of professionals to provide adequate planning and feasibility studies. Many owners do not maintain an in-house engineering and construction management capability, and they should consider the establishment of an ongoing relationship with outside consultants in order to respond quickly to requests. Even among those owners who maintain engineering and construction divisions, many treat these divisions as reimbursable, independent organizations. Such an arrangement should not discourage their legitimate use as false economies in reimbursable costs from such divisions can indeed be very costly to the overall organization.

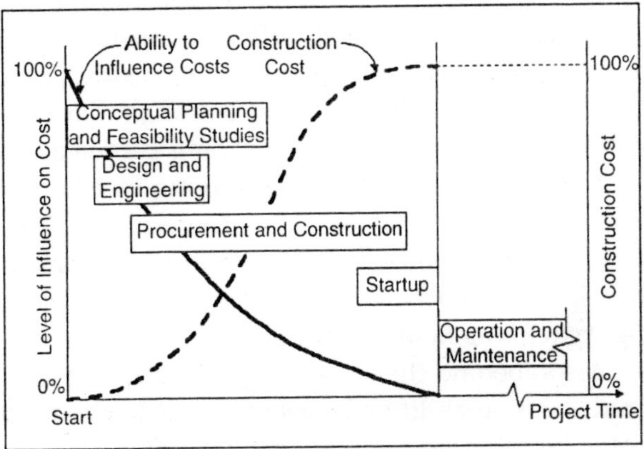

Fig. 2.3 Ability to Influence Construction Cost Over Time

Finally, the initiation and execution of capital projects places demands on the resources of the owner and the professionals and contractors to be engaged by the owner. For very large projects, it may bid up the price of engineering services as well as the costs of materials and equipment and the contract prices of all types. Consequently, such factors should be taken into consideration in determining the timing of a project.

SETTING PRIORITIES FOR PROJECTS

A department store planned to expand its operation by acquiring 20 acres of land in the southeast of a metropolitan area which consists of well established suburbs for middle income families. An architectural/engineering (A/E) firm was engaged to design a shopping center on the 20-acre plot with the department store as its flagship plus a large number of storefronts for tenants. One year later, the department store owner purchased 2,000 acres of farm land in the northwest outskirts of the same metropolitan area and designated 20 acres of this land for a shopping center. The A/E firm was again engaged to design a shopping center at this new location.

The A/E firm was kept completely in the dark while the

assemblage of the 2,000 acres of land in the northwest quietly took place. When the plans and specifications for the southeast shopping center were completed, the owner informed the A/E firm that it would not proceed with the construction of the southeast shopping center for the time being. Instead, the owner urged the A/E firm to produce a new set of similar plans and specifications for the northwest shopping center as soon as possible, even at the sacrifice of cost saving measures. When the plans and specifications for the northwest shopping center were ready, the owner immediately authorized its construction. However, it took another three years before the southeast shopping center was finally built.

The reason behind the change of plan was that the owner discovered the availability of the farm land in the northwest which could be developed into residential real estate properties for upper middle income families. The immediate construction of the northwest shopping center would make the land development parcels more attractive to home buyers. Thus, the owner was able to recoup enough cash flow in three years to construct the southeast shopping center in addition to financing the construction of the northeast shopping center, as well as the land development in its vicinity. While the owner did not want the construction cost of the northwest shopping center to run wild, it apparently was satisfied with the cost estimate based on the detailed plans of the southeast shopping center. Thus, the owner had a general idea of what the construction cost of the northwest shopping center would be, and did not wish to wait for a more refined cost estimate until the detailed plans for that center were ready. To the owner, the timeliness of completing the construction of the northwest shopping center was far more important than reducing the construction cost in fulfilling its investment objectives.

RESOURCE CONSTRAINTS FOR MEGA PROJECTS

A major problem with mega projects is the severe strain placed on the environment, particularly on the resources in the immediate area of a construction project. "Mega" or "macro" projects involve construction of very large facilities such as the Alaska pipeline constructed in the 1970's or the Panama Canal constructed in the 1900's.

The limitations in some or all of the basic elements required for the successful completion of a mega project include:

- Engineering design professionals to provide sufficient manpower to complete the design within a reasonable time limit.
- Construction supervisors with capacity and experience to direct large projects.
- The number of construction workers with proper skills to do the work.
- The market to supply materials in sufficient quantities and of required quality on time.
- The ability of the local infrastructure to support the large number of workers over an extended period of time, including housing, transportation and other services.

To compound the problem, mega projects are often constructed in remote environments away from major population centers and subject to severe climate conditions. Consequently, special features of each mega project must be evaluated carefully.

EFFECTS OF PROJECT RISKS ON ORGANIZATION

The uncertainty in undertaking a construction project comes from many sources and often involves many participants in the project. Since each participant tries to minimize its own risk, the conflicts among various participants can be detrimental to the project. Only the owner has the power to moderate such conflicts as it alone holds the key to risk assignment through proper contractual relations with other participants. Failure to recognize this responsibility by the owner often leads to undesirable results. In recent years, the concept of "risk sharing/ risk assignment" contracts has gained acceptance by the federal government. Since this type of contract acknowledges the responsibilities of the owners, the contract prices are expected to be lower than those in which all risks are assigned to contractors. In approaching the problem of uncertainty, it is important to recognize that incentives must be provided if any of the participants is expected to take a greater risk. The willingness of a participant to accept risks often reflects the professional competence of that participant as well as its propensity to risk. However, society's perception of the potential liabilities of the participant can affect the attitude of risk-taking for all participants. When a claim is made against one of the participants, it is difficult for the public to know whether a fraud has been committed, or simply that an accident has occurred. Risks in construction projects may be classified in a number of ways.

One form of classification is as follows:

- *Socioeconomic factors:*
  - Environmental protection
  - Public safety regulation
  - Economic instability
  - Exchange rate fluctuation
- *Organizational relationships:*
  - Contractual relations
  - Attitudes of participants
  - Communication
- *Technological problems:*
  - Design assumptions
  - Site conditions
  - Construction procedures
  - Construction occupational safety

The environmental protection movement has contributed to the uncertainty for construction because of the inability to know what will be required and how long it will take to obtain approval from the regulatory agencies. The requirements of continued re-evaluation of problems and the lack of definitive criteria which are practical have also resulted in added costs. Public safety regulations have similar effects, which have been most noticeable in the energy field involving nuclear power plants and coal mining. The situation has created constantly shifting guidelines for engineers, constructors and owners as projects move through the stages of planning to construction. These moving targets add a significant new dimension of uncertainty which can make it virtually impossible to schedule and complete work at budgeted cost. Economic conditions of the past decade have further reinforced the climate of uncertainty with high inflation and interest rates. The deregulation of financial institutions has also generated unanticipated problems related to the financing of construction.

Uncertainty stemming from regulatory agencies, environmental issues and financial aspects of construction should be at least mitigated or ideally eliminated. Owners are keenly interested in achieving some form of breakthrough that will lower the costs of projects and mitigate or eliminate lengthy delays. Such breakthroughs are seldom planned. Generally, they happen when the right conditions exist, such as when innovation is permitted or when a basis for incentive or reward exists. However, there is a long way to go before a true partnership of all parties involved can be forged.

During periods of economic expansion, major capital expenditures are made by industries and bid up the cost of construction. In order to control costs, some owners attempt to use fixed price contracts so that the risks of unforeseen contingencies related to an overheated economy are passed on to contractors. However, contractors will raise their prices to compensate for the additional risks.

The risks related to organizational relationships may appear to be unnecessary but are quite real. Strained relationships may develop between various organizations involved in the design/construct process. When problems occur, discussions often center on responsibilities rather than project needs at a time when the focus should be on solving the problems. Cooperation and communication between the parties are discouraged for fear of the effects of impending litigation. This barrier to communication results from the ill-conceived notion that uncertainties resulting from technological problems can be eliminated by appropriate contract terms. The net result has been an increase in the costs of constructed facilities. The risks related to technological problems are familiar to the design/construct professions which have some degree of control over this category. However, because of rapid advances in new technologies which present new problems to designers and constructors, technological risk has become greater in many instances.

Certain design assumptions which have served the professions well in the past may become obsolete in dealing with new types of facilities which may have greater complexity or scale or both. Site conditions, particularly subsurface conditions which always present some degree of uncertainty, can create an even greater degree of uncertainty for facilities with heretofore unknown characteristics during operation. Because construction procedures may not have been fully anticipated, the design may have to be modified after construction has begun. An example of facilities which have encountered such uncertainty is the nuclear power plant, and many owners, designers and contractors have suffered for undertaking such projects.

If each of the problems cited above can cause uncertainty, the combination of such problems is often regarded by all parties as being out of control and inherently risky. Thus, the issue of liability has taken on major proportions and has influenced the practices of engineers and constructors, who in turn have influenced the actions of the owners.

Many owners have begun to understand the problems of risks and are seeking to address some of these problems. For example, some owners are turning to those organizations that offer complete capabilities in planning, design, and construction, and tend to avoid breaking the project into major components to be undertaken individually by specialty participants. Proper coordination throughout the project duration and good organizational communication can avoid delays and costs resulting from fragmentation of services, even though the components from various services are eventually integrated. Attitudes of cooperation can be readily applied to the private sector, but only in special circumstances can they be applied to the public sector. The ability to deal with complex issues is often precluded in the competitive bidding which is usually required in the public sector. The situation becomes more difficult with the proliferation of regulatory requirements and resulting delays in design and construction while awaiting approvals from government officials who do not participate in the risks of the project.

ORGANIZATION OF PROJECT PARTICIPANTS

The top management of the owner sets the overall policy and selects the appropriate organization to take charge of a proposed project. Its policy will dictate how the project life cycle is divided among organizations and which professionals should be engaged. Decisions by the top management of the owner will also influence the organization to be adopted for project management. In general, there are many ways to decompose a project into stages.

The most typical ways are:

- Sequential processing whereby the project is divided into separate stages and each stage is carried out successively in sequence.
- Parallel processing whereby the project is divided into independent parts such that all stages are carried out simultaneously.
- Staggered processing whereby the stages may be overlapping, such as the use of phased design-construct procedures for fast track operation.

It should be pointed out that some decomposition may work out better than others, depending on the circumstances. In any case, the prevalence of decomposition makes the subsequent integration particularly important.

The critical issues involved in organization for project management are:

- How many organizations are involved?
- What are the relationships among the organizations?
- When are the various organizations brought into the project?

There are two basic approaches to organize for project

implementation, even though many variations may exist as a result of different contractual relationships adopted by the owner and builder.

These basic approaches are divided along the following lines:

- *Separation of organizations*: Numerous organizations serve as consultants or contractors to the owner, with different organizations handling design and construction functions. Typical examples which involve different degrees of separation are:
  - Traditional sequence of design and construction
  - Professional construction management
- *Integration of organizations*: A single or joint venture consisting of a number of organizations with a single command undertakes both design and construction functions. Two extremes may be cited as examples:
  - Owner-builder operation in which all work will be handled in house by force account.
  - Turnkey operation in which all work is contracted to a vendor which is responsible for delivering the completed project

Since construction projects may be managed by a spectrum of participants in a variety of combinations, the organization for the management of such projects may vary from case to case. On one extreme, each project may be staffed by existing personnel in the functional divisions of the organization on an ad-hoc basis as shown in Figure 2.4 until the project is completed. This arrangement is referred to as the matrix organization as each project manager must negotiate all resources for the project from the existing organizational framework. On the other hand, the organization may consist of a small central functional staff for the exclusive purpose of supporting various projects,. each of which has its functional divisions as shown in Figure 2.5. This decentralized set-up is referred to as the project oriented organization as each project manager has autonomy in managing the project. There are many variations of management style between these two extremes, depending on the objectives of the organization and the nature of the construction project. For example, a large chemical company with in-house staff for planning, design and construction of facilities for new product lines will naturally adopt the matrix organization. On the other hand, a construction company whose existence depends entirely on the management of certain types of construction projects may find the project-oriented organization particularly attractive. While organizations may differ, the same basic principles of management structure are applicable to most situations.

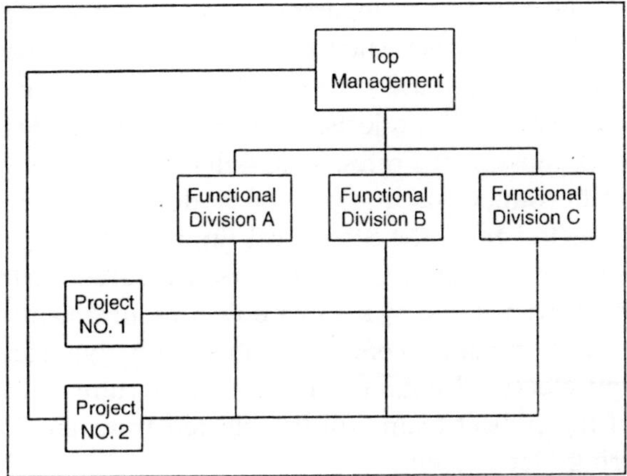

Fig. 2.4 A Matrix Organization

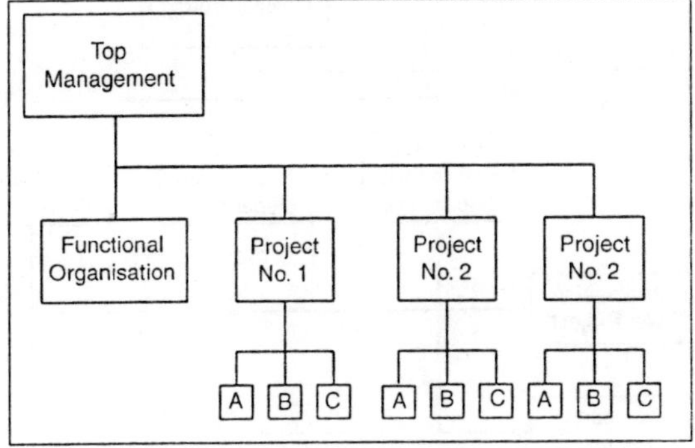

Fig. 2.5 A Project-Oriented Organization

To illustrate various types of organizations for project management, we shall consider two examples, the first one representing an owner organization while the second one representing the organization of a construction management consultant under the direct supervision of the owner.

MATRIX ORGANIZATION OF AN ENGINEERING DIVISION

The Engineering Division of an Electric Power and Light Company has functional departments as shown in Figure 2.6. When small scale projects such as the addition of a transmission tower or a sub-station are authorized, a matrix organization is used to carry out such projects. For example, in the design of a transmission tower, the professional skill of a structural engineer is most important. Consequently, the leader of the project team will be selected from the Structural Engineering Department while the remaining team members are selected from all departments as dictated by the manpower requirements. On the other hand, in the design of a new sub-station, the professional skill of an electrical engineer is most important. Hence, the leader of the project team will be selected from the Electrical Engineering Department.

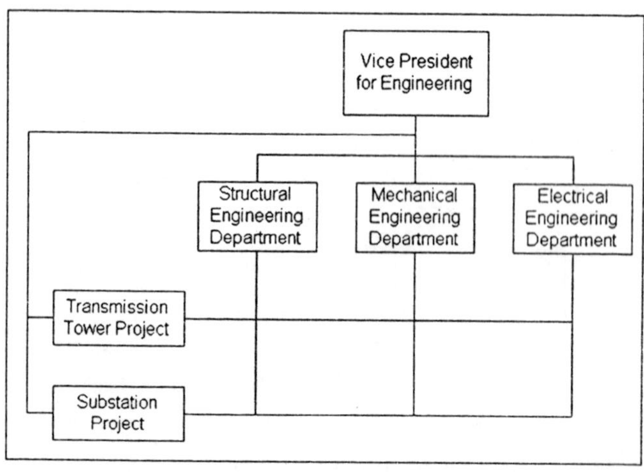

Fig. 2.6 The Matrix Organization in an Engineering Division

CONSTRUCTION MANAGEMENT CONSULTANT ORGANIZATION

When the same Electric Power and Light Company in the previous example decided to build a new nuclear power plant, it engaged a construction management consultant to take charge of the design and construction completely. However, the company also assigned a project team to coordinate with the construction management consultant as shown in Figure 2.7.

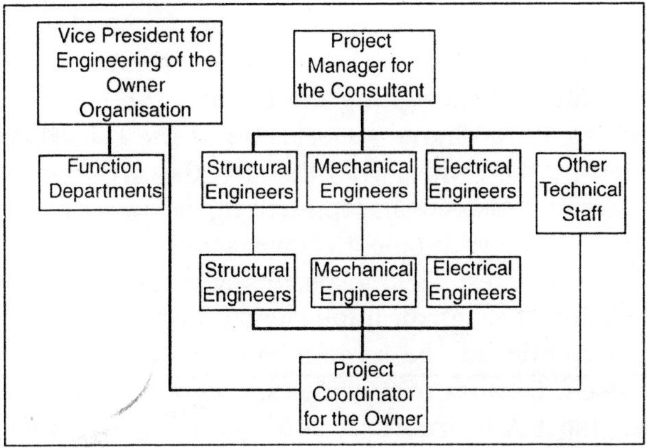

Fig. 2.7 Coordination between Owner and Consultant

Since the company eventually will operate the power plant upon its completion, it is highly important for its staff to monitor the design and construction of the plant. Such coordination allows the owner not only to assure the quality of construction but also to be familiar with the design to facilitate future operation and maintenance.

Note the close direct relationships of various departments of the owner and the consultant. Since the project will last for many years before its completion, the staff members assigned to the project team are not expected to rejoin the Engineering Department but will probably be involved in the future operation of the new plant. Thus, the project team can act independently towards its designated mission.

TRADITIONAL DESIGNER-CONSTRUCTOR SEQUENCE

For ordinary projects of moderate size and complexity, the owner often employs a designer which prepares the detailed plans and specifications for the constructor. The designer also acts on behalf of the owner to oversee the project implementation during construction. The general contractor is responsible for the construction itself even though the work may actually be undertaken by a number of specialty subcontractors.

The owner usually negotiates the fee for service with the architectural/engineering (A/E) firm. In addition to the responsibilities of designing the facility, the A/E firm also exercises to some degree supervision of the construction as stipulated by the owner. Traditionally, the A/E firm regards itself as design professionals representing the owner who should not communicate with potential contractors to avoid collusion or conflict of interest. Field inspectors working for an A/E firm usually follow through the implementation of a project after the design is completed and seldom have extensive input in the design itself. Because of the litigation climate in the last two decades, most A/E firms only provide observers rather than inspectors in the field. Even the shop drawings of fabrication or construction schemes submitted by the contractors for approval are reviewed with a disclaimer of responsibility by the A/E firms.

The owner may select a general constructor either through competitive bidding or through negotiation. Public agencies are required to use the competitive bidding mode, while private organizations may choose either mode of operation. In using competitive bidding, the owner is forced to use the designer-constructor sequence since detailed plans and specifications must be ready before inviting bidders to submit their bids. If the owner chooses to use a negotiated contract, it is free to use phased construction if it so desires.

The general contractor may choose to perform all or part of the construction work, or act only as a manager by subcontracting all the construction to subcontractors. The general contractor may also select the subcontractors through competitive bidding or negotiated contracts. The general contractor may ask a number of subcontractors to quote prices for the subcontracts before submitting its bid to the owner. However, the subcontractors often cannot force the winning general contractor to use them on the project. This situation may lead to practices known as *bid shopping* and *bid peddling*. Bid shopping refers to the situation when the general contractor approaches subcontractors other than those whose quoted prices were used in the winning contract in order to seek lower priced subcontracts. Bid peddling refers to the actions of subcontractors who offer lower priced subcontracts to the winning general subcontractors in order to dislodge the subcontractors who originally quoted prices to the general contractor prior to its bid submittal. In both cases, the quality of construction may be sacrificed, and some state statutes forbid these practices for public projects.

Although the designer-constructor sequence is still widely used because of the public perception of fairness in competitive bidding, many private owners recognize the disadvantages of using this approach when the project is large and complex and when market pressures require a shorter project duration than that which can be accomplished by using this traditional method.

PROFESSIONAL CONSTRUCTION MANAGEMENT

Professional construction management refers to a project management team consisting of a professional construction manager and other participants who will carry out the tasks of project planning, design and construction in an integrated manner. Contractual relationships among members of the team are intended to minimize adversarial relationships and contribute to greater response within the management group.

A professional construction manager is a firm specialized in the practice of professional construction management which includes:

- Work with owner and the A/E firms from the beginning and make recommendations on design improvements, construction technology, schedules and construction economy.

- Propose design and construction alternatives if appropriate, and analyse the effects of the alternatives on the project cost and schedule.
- Monitor subsequent development of the project in order that these targets are not exceeded without the knowledge of the owner.
- Coordinate procurement of material and equipment and the work of all construction contractors, and monthly payments to contractors, changes, claims and inspection for conforming design requirements.
- Perform other project related services as required by owners.

Professional construction management is usually used when a project is very large or complex.

The organizational features that are characteristics of mega-projects can be summarized as follows:

- The overall organizational approach for the project will change as the project advances. The "functional" organization may change to a "matrix" which may change to a "project" organization.
- Within the overall organization, there will probably be functional, project, and matrix suborganizations all at the same time. This feature greatly complicates the theory and the practice of management, yet is essential for overall cost effectiveness.
- Successful giant, complex organizations usually have a strong matrix-type sub organization at the level where basic cost and schedule control responsibility is assigned. This sub organization is referred to as a "cost center" or as a "project" and is headed by a project manager. The cost center matrix may have participants assigned from many different functional groups. In turn, these functional groups may have technical reporting responsibilities to several different and higher tiers in the organization. The key to a cost

effective effort is the development of this project sub organization into a single team under the leadership of a strong project manager.

- The extent to which decision-making will be centralized or decentralized is crucial to the organization of the mega-project.

Consequently, it is important to recognize the changing nature of the organizational structure as a project is carried out in various stages.

MANAGING OF THE ALASKA PIPELINE PROJECT

The Alaska Pipeline Project was the largest, most expensive private construction project in the 1970's, which encompassed 800 miles, thousands of employees, and 10 billion dollars.

At the planning stage, the owner (a consortium) employed a Construction Management Contractor (CMC) to direct the pipeline portion, but retained centralized decision making to assure single direction and to integrate the effort of the CMC with the pump stations and the terminals performed by another contractor. The CMC also centralized its decision making in directing over 400 subcontractors and thousands of vendors. Because there were 19 different construction camps and hundreds of different construction sites, this centralization caused delays in decision making.

At about the 15% point of physical completion, the owner decided to reorganize the decision making process and change the role of the CMC. The new organization was a combination of owner and CMC personnel assigned within an integrated organization. The objective was to develop a single project team responsible for controlling all subcontractors. Instead of having nine tiers of organization from the General Manager of the CMC to the subcontractors, the new organization had only four tiers from the Senior Project Manager of the owner to subcontractors. Besides unified direction and coordination, this reduction in tiers of organization greatly improved communications and the ability to make and implement decisions. The new organization also allowed decentralization of decision making by treating five sections of the pipeline at different geographic locations as separate projects, with a section manager responsible for all functions of the section as a profit center.

At about 98% point of physical completion, all remaining activities were to be consolidated to identify single bottom-line responsibility, to reduce duplication in management staff, and to unify coordination of remaining work. Thus, the project was first handled by separate organizations but later was run by an integrated organization with decentralized profit centers. Finally, the organization in effect became small and was ready to be phased out of operation.

MANAGING THE CHANNEL TUNNEL CONSTRUCTION

The underground railroad tunnel from Britain to France is commonly called the Channel Tunnel or Chunnel. It was built by tunneling from each side. Management turmoil dogged the project from the start. In 1989, seven of the eight top people in the construction organization left.

There was a built in conflict between the contractors and government overseers: "The fundamental thing wrong is that the constractors own less than 6% of Eurotunnel. Their interest is to build and sell the project at a profit. (Eurotunnel's) interest is for it to operate economically, safely and reliably for the next 50 years.".

OWNER-BUILDER OPERATION

In this approach an owner must have a steady flow of on-going projects in order to maintain a large work force for in-house operation. However, the owner may choose to subcontract a substantial portion of the project to outside consultants and contractors for both design and construction, even though it retains centralized decision making to integrate all efforts in project implementation.

The District Engineer's Office of the U.S. Army Corps of Engineers may be viewed as a typical example of an owner-builder approach as shown in Figure 2.8.

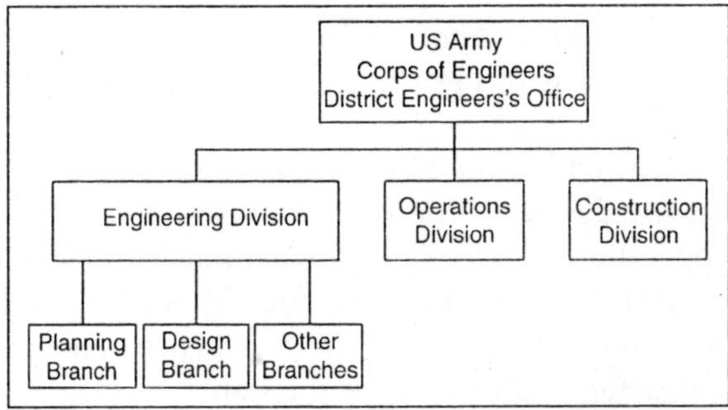

Fig. 2.8 Organization of a District of Corps of Engineers

In the District Engineer's Office of the U.S. Corps of Engineers, there usually exist an Engineering Division and an Operations Division, and, in a large district, a Construction Division. Under each division, there are several branches. Since the authorization of a project is usually initiated by the U.S. Congress, the planning and design functions are separated in order to facilitate operations. Since the authorization of the feasibility study of a project may precede the authorization of the design by many years, each stage can best be handled by a different branch in the Engineering Division. If construction is ultimately authorized, the work may be handled by the Construction Division or by outside contractors. The Operations Division handles the operation of locks and other facilities which require routine attention and maintenance.

When a project is authorized, a project manager is selected from the most appropriate branch to head the project, together with a group of staff drawn from various branches to form the project team. When the project is completed, all members of the team including the project manager will return to their regular posts in various branches and divisions until the next project assignment. Thus, a matrix organization is used in managing each project.

TURNKEY OPERATION

Some owners wish to delegate all responsibilities of design and construction to outside consultants in a *turnkey* project arrangement. A contractor agrees to provide the completed facility on the basis of performance specifications set forth by the owner.

The contractor may even assume the responsibility of operating the project if the owner so desires. In order for a turnkey operation to succeed, the owner must be able to provide a set of unambiguous performance specifications to the contractor and must have complete confidence in the capability of the contractor to carry out the mission.

This approach is the direct opposite of the owner-builder approach in which the owner wishes to retain the maximum amount of control for the design-construction process.

EXAMPLE OF A TURNKEY ORGANIZATION

A 150-Mw power plant was proposed in 1985 by the Texas-New Mexico Power Company of Fort Worth, Texas, which would make use of the turnkey operation. Upon approval by the Texas Utility Commission, a consortium consisting of H.B. Zachry Co., Westinghouse Electric Co., and Combustion Engineering, Inc. would design, build and finance the power plant for completion in 1990 for an estimated construction cost of $200 million in 1990 dollars.

The consortium would assume total liability during construction, including debt service costs, and thereby eliminate the risks of cost escalation to rate payers, stockholders and the utility company management.

LEADERSHIP AND MOTIVATION FOR THE PROJECT TEAM

The project manager, in the broadest sense of the term, is the most important person for the success or failure of a project. The project manager is responsible for planning, organizing and controlling the project. In turn, the project manager receives authority from the management of the organization to mobilize the necessary resources to complete a project.

The project manager must be able to exert interpersonal influence in order to lead the project team.

The project manager often gains the support of his/her team through a combination of the following:

- Formal authority resulting from an official capacity which is empowered to issue orders.
- Reward and/or penalty power resulting from his/her capacity to dispense directly or indirectly valued organization rewards or penalties.
- Expert power when the project manager is perceived as possessing special knowledge or expertise for the job.
- Attractive power because the project manager has a personality or other characteristics to convince others.

In a matrix organization, the members of the functional departments may be accustomed to a single reporting line in a hierarchical structure, but the project manager coordinates the activities of the team members drawn from functional departments. The functional structure within the matrix organization is responsible for priorities, coordination, administration and final decisions pertaining to project implementation. Thus, there are potential conflicts between functional divisions and project teams. The project manager must be given the responsibility and authority to resolve various conflicts such that the established project policy and quality standards will not be jeopardized. When contending issues of a more fundamental nature are developed, they must be brought to the attention of a high level in the management and be resolved expeditiously.

In general, the project manager's authority must be clearly documented as well as defined, particularly in a matrix organization where the functional division managers often retain certain authority over the personnel temporarily assigned to a project.

The following principles should be observed:

- The interface between the project manager and the functional division managers should be kept as simple as possible.
- The project manager must gain control over those

elements of the project which may overlap with functional division managers.

- The project manager should encourage problem solving rather than role playing of team members drawn from various functional divisions.

INTERPERSONAL BEHAVIOUR IN PROJECT ORGANIZATIONS

While a successful project manager must be a good leader, other members of the project team must also learn to work together, whether they are assembled from different divisions of the same organization or even from different organizations. Some problems of interaction may arise initially when the team members are unfamiliar with their own roles in the project team, particularly for a large and complex project. These problems must be resolved quickly in order to develop an effective, functioning team.

Many of the major issues in construction projects require effective interventions by individuals, groups and organizations. The fundamental challenge is to enhance communication among individuals, groups and organizations so that obstacles in the way of improving interpersonal relations may be removed. Some behaviour science concepts are helpful in overcoming communication difficulties that block cooperation and coordination. In very large projects, professional behaviour scientists may be necessary in diagnosing the problems and advising the personnel working on the project. The power of the organization should be used judiciously in resolving conflicts.

The major symptoms of interpersonal behaviour problems can be detected by experienced observers, and they are often the sources of serious communication difficulties among participants in a project. For example, members of a project team may avoid each other and withdraw from active interactions about differences that need to be dealt with. They may attempt to criticize and blame other individuals or groups when things go wrong. They may resent suggestions for improvement, and become defensive to minimize culpability rather than take the initiative to maximize achievements. All these actions are detrimental to the project organization.

While these symptoms can occur to individuals at any organization, they are compounded if the project team consists of individuals who are put together from different organizations. Invariably, different organizations have different cultures or modes of operation. Individuals from different groups may not have a common loyalty and may prefer to expand their energy in the directions most advantageous to themselves instead of the project team. Therefore, no one should take it for granted that a project team will work together harmoniously just because its members are placed physically together in one location. On the contrary, it must be assumed that good communication can be achieved only through the deliberate effort of the top management of each organization contributing to the joint venture.

PERCEPTIONS OF OWNERS AND CONTRACTORS

Although owners and contractors may have different perceptions on project management for construction, they have a common interest in creating an environment leading to successful projects in which performance quality, completion time and final costs are within prescribed limits and tolerances. It is interesting therefore to note the opinions of some leading contractors and owners who were interviewed in 1984.

From the responses of six contractors, the key factors cited for successful projects are:

- Well defined scope
- Extensive early planning
- Good leadership, management and first line supervision
- Positive client relationship with client involvement
- Proper project team chemistry
- Quick response to changes
- Engineering managers concerned with the total project, not just the engineering elements.

Conversely, the key factors cited for unsuccessful projects are:

- Ill-defined scope
- Poor management
- Poor planning
- Breakdown in communication between engineering and construction
- Unrealistic scope, schedules and budgets
- Many changes at various stages of progress
- Lack of good project control

The responses of eight owners indicated that they did not always understand the concerns of the contractors although they generally agreed with some of the key factors for successful and unsuccessful projects cited by the contractors.

The significant findings of the interviews with owners are summarized as follows:

- All owners have the same perception of their own role, but they differ significantly in assuming that role in practice.
- The owners also differ dramatically in the amount of early planning and in providing information in bid packages.
- There is a trend towards breaking a project into several smaller projects as the projects become larger and more complex.
- Most owners recognize the importance of schedule, but they adopt different requirements in controlling the schedule.
- All agree that people are the key to project success.

From the results of these interviews, it is obvious that owners must be more aware and involved in the process in order to generate favorable conditions for successful projects. Design professionals and construction contractors must provide better communication with each other and with the owner in project implementation.

3

Regulations and Safety

CONSTRUCTION REGULATIONS

SUBCONTRACTOR SAFETY POLICY

General

Lawrence Berkeley National Laboratory (LBNL) is committed to providing and maintaining the safest possible work conditions for all workers by promoting the integration of safety management into all construction processes. Project managers, construction managers, superintendents, assistant/area superintendents, safety representatives, and foremen are responsible for implementing and maintaining an effective safety programme. It is the responsibility of these individuals to ensure workers under their supervision maintain safe work areas and perform their tasks in a safe manner. It is also the responsibility of each worker to follow every precaution and LBNL safety rule and policy to protect themselves and their fellow workers. LBNL will monitor subcontractor safety programmes and performance to ensure compliance with LBNL requirements. LBNL safety audits will be available to the LBNL Procurement Office for evaluation of the subcontractor's safety performance.

When performing work at LBNL, each subcontractor:

- Is responsible for the safety of their employees and/or visitors as required by the rules and regulations of this chapter; 10 CFR 851 "Worker Safety and Health Programme"; the California Code of Regulations, Title 8, Construction Safety Orders; 29 CFR 1926 Safety Standards for the Construction Industry; and all other local, state, and federally recognized current standards and codes.
- Is required to understand and follow the contents of this chapter.
- Is responsible for training and educating their employees, and/or visitors, as to the contents of this chapter and requirements for conduct of work under the LBNL Integrated Safety Management (ISM) Plan. (Documentation of all training is the responsibility of the subcontractor.)

Where conflict between cited standards or safe practices occur, the EHS Division will determine which standard shall apply.

SCOPE

GENERAL

This policy applies to all construction activities performed by LBNL employees; contract employees under the direct technical supervision of an LBNL employee, referred to as "contract employees"; and non-Laboratory employees under the supervision of subcontractors, referred to as "construction subcontractors." Construction activities may originate from construction subcontracts, service contracts, purchase orders, and in-house work orders.

Department of Energy 10 CFR 851, "Worker Safety and Health Programme"

The subcontractor shall comply with the Department of Energy's Worker Safety and Health Programme regulation, 10 CFR 851, which enforces worker safety and health requirements including but not limited to standards of the Occupational Safety and Health Administration (OSHA) as incorporated in the LBNL Worker Safety and Health Programme. Violations of safety and health provisions of 10 CFR 851 may subject subcontractor to penalties. The subcontractor shall follow the provisions of its Cal/OSHA mandated Injury and Illness Prevention Plan (IIPP) and submit it to LBNL when requested. The subcontractor shall also follow all LBNL safety procedures and policies communicated in it.

Contract Employees

The safety rights and obligations of contract employees are the same as those of Laboratory employees. Those supervisors assigned to direct the work of contract employees must verify that contract employees are afforded conditions equivalent to those provided to Laboratory employees, including training and personal protective equipment (PPE).

CONSTRUCTION SUBCONTRACTOR SELECTION PROCESS

Construction subcontractor selection shall include an evaluation of the subcontractor's prior safety performance, the subcontractor's current written safety programmes, and qualifications of key EH&S personnel to assure LBNL that the subcontractor is capable of meeting safety performance goals. All subcontractors being considered to perform construction activities at LBNL-controlled sites and property shall undergo such an evaluation.

During the construction subcontractor selection process, Facilities Division Procurement personnel request evidence of the subcontractor's workers' compensation experience modification rates (EMRs). In general, the subcontractor must achieve an EMR of 1.0 or less to be considered for a contract award. Occasionally, a subcontractor's EMR may not be an accurate representation of safety performance (due to statistical variations in the EMR calculation from small payroll numbers, outlier workers' compensation events, etc.). In these cases, an evaluation of a minimum of the last three years of the subcontractor's OSHA Form 300 and 300A data may be used to qualify a subcontractor from a construction safety standpoint.

Additionally, during the selection process a copy of the subcontractor's Cal/OSHA-required Injury and Illness Prevention Programme (IIPP) is requested for review. The IIPP is a written programme in which the subcontractor details the means and methods used to ensure the safety and health of the subcontractor's employees.

The LBNL Project Management Team and Procurement personnel shall evaluate the information submitted by subcontractors. Subcontractors shall demonstrate an equivalent process in the selection of lower-tier subcontractors and shall submit evaluation criteria and results to LBNL.

SUBCONTRACTOR ON-SITE HEALTH AND SAFETY REPRESENTATIVE QUALIFICATIONS

The subcontractor shall provide a qualified on-site Health and Safety Representative, accepted by the LBNL Project Manager, with the authority to enforce all of the safety requirements of this programme, including implementation of the subcontractor's Injury and Illness Prevention Programme and Project EH&S Plan.

LBNL project management and EH&S will make a risk-based decision as to the qualification level of the subcontractor EH&S representative. Requirements may range from a full-time on-site safety professional (Certified Safety Professional) to a craft supervisor with competency as measured by experience training. During periods of active construction (*i.e.*, excluding weekends, weather delays, or other periods of work inactivity), the subcontractor must have a designated representative on the construction worksite who is knowledgeable of the project's hazards and has full authority to act on behalf of the subcontractor. The subcontractor's designated representative must make frequent and regular inspections of the construction worksite to identify and correct any instances of noncompliance with the project health and safety requirements.

Qualification Evaluation

Based on the level of EH&S qualification determined necessary by LBNL, the subcontractor shall submit the following documentation, for review and acceptance by the LBNL Project Manager, in support of the proposed candidate:

- Professional certifications (CSP, CIH, ASP, etc.).
- Curriculum vitae detailing work experience and EH&S responsibilities on projects of similar scope for the previous five years, at a minimum.
- Evidence of construction safety training with a minimum of the 10-hour OSHA training (the requirement will be modified as necessary and may require the 30-hour OSHA construction supervision training or successful completion of the OSHA 500 programme).
- Proof of Competent Person or Qualified Person status attained by the proposed on-site EH&S representative.

A subcontractor shall replace his or her EH&S Representative at the discretion of the LBNL Project Manager within 24 hours upon written notification if the EH&S Representative is unsuccessful in enforcing project safety requirements.

Subcontractor Health and Safety Representative Responsibilities

The subcontractor EH&S Representative shall:

- Assist in the development of the subcontractor's safety plan and job site management system
- Support training of subcontractor personnel
- Evaluate the subcontractor safety process continuously
- Respond to questions regarding the subcontractor safety process
- Attend pre-job meetings to discuss their site-specific safety plan Conduct and document job site safety audits.
- Assist in the identification of jobs requiring an activity-based hazard analysis.

Competent Person

Each subcontractor shall provide to LBNL a written list of those persons on-site who are capable of identifying existing and predictable hazards in the surrounding or working conditions which are unsanitary, hazardous, or dangerous to employees, and who has the authorization to take prompt corrective measures to eliminate them.

Subcontractors shall ensure that each competent person listed has been trained in the following areas as applicable:

- Asbestos
- Cranes
- Confined Space
- Demolition
- Excavations
- Fall Protection
- Ladder
- Scaffold
- Steel Erection
- Underground Construction

Qualified Person

Each subcontractor shall provide to LBNL a written list of those persons on-site who by possession of a recognized degree, certification, or professional standing, or who by extensive knowledge, training, and experience, has successfully demonstrated their ability to solve or resolve problems relating to the subject matter, the work, or the project.

Subcontractors shall ensure that each qualified person listed has been trained in the following areas as applicable:

- Asbestos
- Hazardous Waste Operation
- Crane
- Chemical Hazards
- Concrete and Masonry Const.
- Steel Erection, Open Joists, Site Layout, and Plan
- Hoist and Rigging
- Scaffold

- Underground Const.
- Gases, Vapours, Dust, Mist,
- Ventilation
- Material Hoisting, Personnel Hoist, Elevators
- Excavation

SUBCONTRACTOR INJURY AND ILLNESS PREVENTION PROGRAMME (IIPP)

Each subcontractor shall provide LBNL with a written IIPP. *The IIPP or the IIPP with a Site Specific Safety Plan shall contain at a minimum*:

- The name of the On-site Health and Safety Representative who is responsible for the implementation of the plan their qualifications and what roles this person will play during the project.
- The company policy and senior company officer statement on environment, safety and health.
- The company safety management and safety oversight plan and correlation with LBNL Integrated Safety Management Plan.
- A list of those project activities for which subsequent activity-based hazard analysis are to be performed.
- The company policy on substance abuse and testing policies if applicable.
- Code of Safe Practices.
- How and when each contactor will conduct their toolbox safety talks.
- Provisions for conducting and documenting job site safety inspections by supervision.
- Training methods used to meet OSHA training requirements to ensure that LBNL Safety Programme requirements are communicated all subcontractor personnel.
- Incident reporting, first aid, and emergency procedures.
- The company safety recognition/incentive policy that will be in effect for this project.
- List of all competent persons overseeing those tasks

in which OSHA requires such person(s), such as excavation, confined space, fall protection, etc.

- All other requirements as in the LBNL Construction Specifications section 01020, EH&S General Requirements as tailored to the project.

Lower tier subcontractors may utilize and abide by the general contractor's written (IIPP) or Site-specific safety programme. General contractors that engage lower tier subcontractors shall submit written documentation for each lower tier subcontractor that demonstrates the methods of compliance with the site specific EH&S plan requirements.

JOB HAZARDS ANALYSIS

Subcontractors shall submit an Activity-Based Hazard Analysis *i.e.*, Job Hazards Analysis (JHA) or Activity Hazard Analysis (AHA) for those construction activities meeting the requirements for performing JHA in section 10.6.1 The JHA shall be approved by the LBNL Project Manager before work commences. Each employee scheduled to work in the activities identified below shall receive safety training in those activities prior to working on them. The subcontractor shall maintain proof of employee training at the work site and make it available to the LBNL Project Manager upon request (Toolbox safety meetings are an acceptable forum to meet this requirement). Subcontractor shall be responsible for submitting a JHA and work procedures to LBNL Project Manger for review a minimum of seven days prior to the start of work for most work activities. The subcontractor must maintain 2 copies of the LBNL approved JHA. The first copy to be maintained with the project files. The second copy shall be posted on the jobsite/project in a conspicuous location.

JHA Requirements

A JHA shall be written based on the following conditions:

- Jobs with the highest injury or illness rates
- Jobs with the potential to cause severe or disabling injuries or illness, even if there is no history of

previous accidents

- Jobs in which one simple human error could lead to a severe accident or injury
- Jobs that are new to your operation or have undergone changes in processes and procedures
- Jobs complex enough to require written instructions.

If not otherwise specified in a particular project specification, the JHA shall be performed in accordance with the OSHA 3071 JHA processes. In general the JHA will include:

- Description of work phase or activity
- Identification of potential hazards associated with the activity
- Address further hazards revealed by supplemental site information (*e.g.*, site characterization data, as-built drawings) provided by the subcontractors construction manager.
- A list of the Subcontractor's planned controls to mitigate the identified hazards
- Identification of specialized training required
- Identification of special permits required
- Name of the Subcontractor's Competent persons responsible for inspecting the activity and ensuring that all proposed safety measures are followed.

Construction activities for which a Job Hazards Analysis may be required include, but are not limited to:

- Roofing
- Hoisting and handling of materials
- Excavations
- Trenching and drilling
- Concrete placement and false work
- Welding
- Steel erection
- Work performed six foot or higher above ground
- Electrical work
- Demolition
- Work in confined spaces
- Work that causes the release of silica such as

demolition or drilling of concrete or work with materials that contain silica

- Work with epoxy coatings
- Work with or around hazardous materials
- Work on hilly terrain
- Use and handling of flammable materials

Additional Requirements for Reporting Hazardous Conditions

Workers must be instructed to report to the subcontractor's designated representative hazards not previously identified or evaluated. If immediate corrective action is not possible or the hazard falls outside of project scope, the construction subcontractor must immediately notify affective workers, post appropriate warning sings, implement needed interim controls measures, and notify the LBNL Construction Manager of the action taken. The subcontractor or the designated representative must stop work in the affected area until appropriate protective measures are established.

Coordination and Tracking of EH&S Construction Safety Package Reviews

The EH&S Division Construction Safety Engineer (CSE) is responsible for coordination, tracking, quality assurance and final approval for the ES&H review of LBNL Construction Projects. The Construction Managers/Project Managers (CM/PMs) receive all required safety documents from the subcontractors and forward them as a package to the CSE. Most safety document packages will require review by several EH&S Subject Matter Experts (SMEs, *i.e.*, Radiation Safety, Environmental, Industrial Safety, Industrial Hygiene, Occupational Medicine, etc.). The CSE reviews the packages and distributes appropriate parts to the appropriate SMEs for their review, comment and approval. The SMEs then return their completed documents to the CSE, who combines comments, identifies and resolves conflicts, and performs a quality review to assure all required submittals are included in the final safety document package. The CSE signs the final package and returns it to CM/PM to begin construction.

Copies of the final safety document package, and review comments are maintained by the CSE for use during construction. When the construction work is completed the safety document package is filed with the Project file.

WORKSITE SAFETY OBSERVATIONS

EH&S Division Construction Safety Engineers and Facilities Division Construction Management conduct routine observations of construction worksites to identify and correct unsafe workplace conditions and behaviors. Both "at-risk" and "safe" conditions and behaviors are identified during the observations. These conditions are recorded in a construction safety observation database that is used to track and report trends in construction safety performance.

DETERMINING CLASSIFICATION OF AT-RISK OBSERVATIONS

Each observation of an at-risk condition or behaviour is classified as De-Minimis, Low, Medium and High and inputted into the construction safety observation database by the observer. Classification of at-risk observations are based on a risk assessment methodology that utilizes a risk assessment table having a 4×4 matrix of impact and probability. The risk assessment table is based on the "Risk Registry Risk Assignment Matrix" found in Appendix P of the LBNL Facilities Division, Construction Projects Department, Project Management Manual.

Table. At-Risk Observation Classification Risk Assessment

	No Injury	**First Aid**	**Medium Severity**	**High Severity**
High Probability	De-Minimis	Medium	High	High
Medium Probability	De-Minimis	Low	Medium	High
Low Probability	De-Minimis	Low	Low	Medium
No Probability	De-Minimis	De-Minimis	De-Minimis	De-Minimis

Construction Safety Observation Relationship to ORPS and NTS Reporting

All Medium and High construction safety at-risk observations are reported to the appropriate responsible individuals for review as possible ORPS and NTS reportable events.

ENGINEERED PROTECTIVE SYSTEMS

Subcontractor shall submit for review to the LBNL's Project Manager, any worker, environment or property protective system required by EH&S regulation to be designed by a registered professional engineer. LBNL's review of such system is solely to verify that the subcontractor has had the required protective systems prepared and stamped by a registered professional engineer.

LBNL's review of any documents showing the design or construction of protective systems for worker and property protection shall not relieve the subcontractor of its obligations to comply with applicable laws and standards for the design and construction of such protective work. Subcontractor shall indemnify and hold harmless LBNL and the Architect Engineer from any and all claims, liability, costs, actions and causes of action arising out of or related to the failure of such protective systems. The subcontractor shall defend LBNL, its officers, employees and agents and the Architect-Engineer in any litigation or proceeding brought with respect to the failure of such protective systems. The cost of required safety engineering services required for safety and protective systems shall be borne solely by the subcontractor and shall be deemed to have been included in the amount bid for the work as stated in the subcontract.

PROCUREMENT OF HAZARDOUS MATERIAL

The subcontractor shall submit to the LBNL Project Manager, for review by the LBNL's EHS Division, any proposed procurement, stocking, installing, or other use of materials containing asbestos, cadmium, chromates, or lead. Additionally, the subcontractor shall submit the product Material Safety Data Sheet (MSDS) for architectural and surface coatings, solvents, adhesives, sealants, oils, compressed gases, pesticides, herbicides, welding materials, or other chemicals used in the construction process, for review and acceptance by the LBNL EHS Division prior to the start of work. All materials and applications shall comply with requirements of any and all of the Bay Area Air Quality Management Districts Regulations, including, but not limited to architectural coatings, general solvent and surface coatings, solvent cleaning operations, adhesive and sealants, visible emissions, and asbestos.

Subcontractor shall keep and maintain proof of compliance with the above-referenced regulations, including any recordkeeping obligations, for a period of two years after completion of the project. Subcontractor shall mal.e such documents or evidence available if so requested by BAAQMD or Lawrence Berkeley National Laboratory.

SAFETY TRAINING AND EDUCATION

Subcontractors shall provide a workforce that is trained to the requirements set forth in general and in the specific substance-and subject-specific standards of 10 CFR 851, 29 CFR 1926, and California Code of Regulations, Title 8, Construction Safety Orders. Subcontractors shall be able to demonstrate satisfaction of training requirements.

SITE ORIENTATION AND PRE-JOB TRAINING

All subcontractors and their lower tier subcontractors shall require and administer pre-job training/orientation on LBNL and subcontractor safety programme requirements to all employees prior to engaging in work activities. The Subcontractor shall maintain on the work site a detailed outline of the orientation and a signed and dated roster of all employees who have completed the project EHS indoctrination.

The orientation programme shall address the following elements at a minimum:

- LBNL Integrated Safety Management Core Functions
- Employee rights and responsibilities
- Responsibility for Stop Work Order (imminent danger)

- Construction subcontractor responsibilities
- Alcohol and drug abuse policy
- Subcontractor's disciplinary procedures
- First aid and medical facilities
- Site and project specific hazards
- Hazard recognition and procedures for reporting or correcting unsafe conditions or practices
- Procedures for reporting accidents and incidents
- Fire fighting and other emergency procedures to include local warning and evacuation systems
- Hazard Communication Programme
- Access to employee exposure monitoring data and medical records
- Protection of the environment, including air, water, and storm drains from construction pollutants
- Location of and access to reviewed project Illness and Injury Prevention Programme, Hazard Analysis and Hazard Abatement Plan
- Location and contents of required posting

WEEKLY SAFETY TOOLBOX TALKS

All subcontractors are required to conduct and document weekly safety toolbox talks. These talks shall be conducted at the site and contain safety information that will increase safety awareness on this project. *The weekly took box talks must relate to the work that is underway or immediately forthcoming.* Attendance by all site personnel is mandatory. An attendance roster of signatures shall be collected and filed with a copy of the toolbox talk. Copies of the attendance roster and toolbox talk shall be forwarded to LBNL upon request. Subcontractors may attend 'All Hands' toolbox talks if a separate list of signatures identifying the subcontractor personnel is maintained. Each subcontractor is responsible for ensuring employee attendance at the Toolbox Safety Talks.

SAFETY INSTRUCTION FOR EMPLOYEES

- When workers are first employees they shall be given

instruction regarding the hazards and safety precautions applicable to the type of work in question and directed to read the Code of Safe Practices.

- The subcontractor shall permit only qualified persons to operate equipment and machinery.
- Where employees are subject to known job site hazards, such as flammable liquids and gases, poisons, caustics, harmful plants and animals, toxic materials, confined spaces, etc., they shall be instructed in the recognition of the hazard, in the procedures for protecting themselves from injury, and in the first aid procedure in the event of injury.

EMERGENCY PROCEDURES

An emergency is any situation that poses an immediate threat to life or property. This would include but not be limited to collapse of a building or a portion thereof, fire, explosion, equipment failure such as collapse of a crane, release or exposure to toxic liquids, vapours, or fumes, presence of gas or other explosive atmospheres, flood, earthquake, etc. Violent or suspicious behaviour may also be cause for initiating emergency procedures.

PROCEDURES

Each subcontractor must ensure that they maintain one person currently qualified in American Red Cross or equivalent CPR and First Aid on site at all times. In the event of a life threatening of other serious incident (fire, injury, etc.) requiring the assistance of outside personnel, contact LBNL emergency services immediately at:

- Call 7-9-1-1 for assistance or from outside line call 9-1-1

Upon calling, the person shall state their name, their contractor's name, the location of the emergency, and the type of emergency.

The caller must stay on the call until released by the emergency dispatcher. The LBNL project Manager shall be contacted as soon as practicable. The LBNL Project Manager will provide direction to the subcontractor on evacuation procedures as the job progresses. For emergencies involving evacuation, all subcontractor personnel shall follow the developed, posted evacuation routes to their designated assembly point, remain there until they are accounted for, and an "all clear" or alternate directive is given. Subcontractors shall disclose designated assembly points for their employees to LBNL. In the event of an evacuation, the subcontractor shall immediately notify LBNL of any missing personnel.

Following the occurrence of an emergency, the contractor shall ensure that all proper incident reports are completed and distributed in the required timeframe. A list of "key" on-site and home office personnel (with 24 hour phone numbers) shall be developed by each subcontractor and submitted to the LBNL Project Manager prior to commencement of work.

The LBNL Project Manager or designated LBNL management team member in off-hours will take charge in the event of a major catastrophe. One or all steps are to be followed:

- Stop work
- Take whatever actions are needed to make people on the project safe.
- Call 7-9-1-1 for assistance or from outside line call 9-1-1
- If necessary, call for site evacuation with role call and clear site access roads
- Issue instructions to all supervisors and employees
- Set up security control at the emergency area
- Refer all media requests to the LBNL Media Affairs Office.

INCIDENT REPORTING

In the event a job site accident occurs, the subcontractor shall immediately implement controls and restrictions on the accident site to ensure the site remains undisturbed until released in- writing by the LBNL Project Manager to resume work.

PROCEDURES

Near Miss/Injury-free Event

It is the responsibility of the subcontractor, to complete all Near Miss investigation, and to report these occurrences with recommendations/implementation of corrective actions. The report will be submitted to the LBNL Project Manger within 24 hours.

First Aid Event

It is the responsibility of the subcontractor to collect and log the contractors' incident reports and recommend corrective action. The incident logs and work hour statistics will be sent to the LBNL Construction Safety Engineer by the 1st of each month.

Medical Treatment Event

If the injury is considered an emergency, immediately call 7- 9-1-1 from Lab phones and 9-1-1 from any cell phone. It is the responsibility of the subcontractor to immediately notify the LBNL Project Manager. The subcontractor shall furnish a copy of their OSHA Form 301(or equivalent) to the LBNL Project Manager within 5 days of the injury.

Serious Injury Event

It is the responsibility of the Prime subcontractor to immediately notify the LBNL Project Manager, of any serious event requiring medical treatment. Refer ALL media inquiries to LBNL Media Relations, Media Affairs Office, at office phone 510-486-7586, or cell phone 510-610-3991.

Fatality

It is the responsibility of the subcontractor to notify the LBNL Project Manager who will then notify the appropriate DOE office. It is the responsibility of the subcontractor to notify Cal/OSHA within 8 hours. Refer ALL media inquiries to the LBNL Management Contact, Media Affairs Office at office phone 510-486-7586, or cell phone 510-610-3991.

STOP WORK ORDER

A stop work order must be given when imminent danger is identified or where significant damage to equipment or property or environmental degradation could occur if the operation continued. When a stop work order is issued, only those areas of a construction project immediately involved in the identified hazardous situation are to be included in the order. Any employee that observes an imminent-danger situation is responsible for stopping the work and reporting it to the subcontractor representative at the work site;

Immediately after stopping work, the person issuing the order must report to the LBNL Project Manager, LBNL Construction Safety Engineer, of his/her action. The LBNL Project Manager, LBNL Construction Safety Engineer will be dispatched to the site to verify that the operation has stopped and that the stop order was exercised in a justifiable and responsible manner.

Work cannot restart until the LBNL Project Manager has agreed that the imminent danger has been eliminated. Notification to restart work will be passed to the construction subcontractor by the LBNL Project Manager.

STOP WORK MEMORANDUM SAFETY DEFICIENCY NOTICE

Following a Stop Work action, a Safety Deficiency Notice will be issued to the subcontractor by LBNL Construction Safety Engineer; it will reference the appropriate OSHA regulation and the subcontractor provision that caused the work stoppage.

DIFFERENCE OF OPINION

Differences of opinion regarding a stop work order between the LBNL Project Manager, LBNL Construction Safety Engineer, and others involved must be immediately referred to their respective functional supervisors for resolution. The recommendations of the LBNL Construction Safety Engineer must be followed until a final decision is made. Final determination will come from the LBNL EHS Division Director.

EMPLOYEE NONCOMPLIANCE TO SAFETY POLICIES

In an effort to ensure compliance to this programme and all other established OSHA standards, LBNL hereby implements this procedure of non compliance to all subcontractors working on LBNL controlled property. This is established to promote safety and eliminate offenders and repeat offenders, and may lead up to contract termination with a subcontractor. This programme may be used or may be superseded with more severe discipline based on the degree of the infraction(s). In any case LBNL Project Team has sole authority in the type of discipline action to include removal from the project.

- First offence, give a verbal warning keep written record of the offence/notify his/her supervisor
- Second offence, give a written warning with his/her supervision is brought into the office for a "discussion" with the subcontractor Superintendent and the LBNL Project Manager along with the LBNL Construction Safety Engineer. A copy of the written warning is sent to the offending Workers Company's office, with a statement to the effect that if this happens again the worker will be removed from the project and could lead to a termination of the contract.
- Third offence the worker is removed from the project.
- If repeat occurrences with other crewmembers are found the supervisor of said offenders shall be subject to removal from the project.

VISITORS

All visitors are required to report to the project field office upon entering the project site. Access to the site shall be denied to any individual who does not have justifiable business on the job site. Requests for tours of the project site shall be carefully screened and limited in frequency and numbers of people. Tours of the site shall be approved by the LBNL Project Manager and LBNL Construction Manager and shall be conducted during non-working hours.

LBNL shall establish the time and travel route for any tour. Areas which may present hazards to the tour groups shall be prohibited. The tour's travel route shall be cleared of any tripping hazards, cleaned, and properly protected to avoid potential personal injury. A designated member of the LBNL management team shall guide the approved tours. All members of a tour group shall sign a release prior to touring the site.

Any project site visitors who are permitted access to the site but are not on official on-site business shall sign the release before being authorized to proceed beyond the project office.

All visitors must wear long pants, shirts with sleeves over the shoulder, hard hats, safety glasses, and hard-soled work shoes or boots when on site. No penny loafers, dress shoes, etc. shall be permitted.

CDM REGULATIONS

Health and safety in the construction industry has experienced considerable research over the last few decades in order to reduce the number of injuries and fatalities within the construction industry. Health and safety practices such as the HSC that introduced the CDM regulations in 1994 and the current revised regulations introduced in 2007 are perceived to provide numerous benefits in maintaining a safer industry. With approximately 7%-8% of employment within the UK involved in the construction industry and an estimated 8% of the UK's GDP it is important to do so.

There is a growing importance due to the high rate of employment in the construction industry to provide a detailed set of safety regulations known as the CDM regulations that everyone involved with construction must follow. As a result of this many organisations such as the HSC, have initiated targets to ensure that the health and safety on sites improves dramatically, and to encourage the government to formulate official targets to reduce the number of fatalities and injuries by a certain date.

However, although the CDM regulations 1994 did have a positive effect on the health and safety in the industry there was still areas of concern within the regulations that professionals were unsure of who carried out different roles and the fact that the rate of fatalities and injuries was not decreasing significantly enough, this was the reason for the review of the CDM regulations 1994 and the implementation of the new CDM regulations 2007 which is designed to provide a clearer background into construction H&S. Using a sample group methodology, and acknowledging specifically the principal contractor, this study aims to assess whether the HSC can achieve its aim of reducing the number of fatalities and injuries by 10% in the UK by 2010 and ultimately assess whether or not the CDM regulations make a significant difference to H&S statistics. This will be reviewed by looking at the implementation techniques used by the PC's on site and develop a framework to provide the best technique in which to do so.

INTRODUCTION

The construction industry has had H&S regulations in place known as the CDM regulations for a number of years now which have recently been renewed. The CDM regulations have application to a majority of construction projects undertaken in the UK which provide the industry with specific standards involving H&S that are considered acceptable to the government and alternatively, to the construction worker.

The latest regulations represent an important trend towards the paternalistic intervention of the government in relation to health and safety. Historically, the ambit of health and safety on site was left to the individual company; this was due to the requirements of the Conservative governments, and if things went wrong it was up to the individual to cope. This tendency has developed during the twentieth century with great attention in recent years of government focus in developing the H&S of the construction worker.

2.2 million People work in Britain's construction industry, making it the country's biggest industry, but it accounts for 16% of major accidents and 25% of fatal injuries. It is has also been acknowledged that for a number of years in the UK it has been one of the most dangerous industries to be employed in. In the last 25 years, over 2,800 people have died from injuries they received as a result of construction work. Many more have been injured or made ill. HSE deals with all aspects of construction work in England, Scotland and Wales. In 2006-2007 alone there were 77 fatal injuries to workers in the construction industry, this was a 28% increase to the 59 in 2005-2006. The 77 fatalities in the construction industry during 2006-2007 accounted for 32% of all worker deaths. The rate of fatal injury to workers per hundred thousand rose to 3.7 in 2006-2007 compared to that of 3.0 in 2005-2006.

Improvements regarding the Health and Safety in the construction industry are urgently needed as it remains a disproportionately dangerous industry which is important not only for the people involved in the industry but also for the expected 2-3% growth within the next 5 years. The improvements require significant and permanent changes in duty holder attitudes and behaviour. Since the original CDM Regulations were introduced in 1994, concerns were raised that their complexity and the bureaucratic approach of many duty holders frustrated the Regulations' underlying health and safety objectives. These views were supported by an industry-wide consultation in 2002 which resulted in the decision to revise the Regulations. The new CDM 2007 Regulations revise and bring together the CDM Regulations 1994 and the Construction Regulations 1996 into a single regulatory package known as the CDM regulations 2007.

The health and safety process has always been a much talked about topic within England and Wales every since the first CDM regulations were launched in 1994. The need to reform this process has been highlighted by research carried out by the health and safety executive which shows that there is need for improvement in the implementation of these regulations which is the main reason for the introduction of the CDM regulations 2007. The research shows that: These have been developed in close consultation with construction industry stakeholders.

The proposal for the CDM 2007 aims to:

- Simplify the Regulations to improve clarity and make it easier for everyone to know what is expected of them;
- Maximise their flexibility to fit with the vast range of contractual arrangements;
- Focus on planning and management, rather than 'the plan' and other paperwork;
- Encourage co-ordination and co-operation, particularly between designers and contractors; and
- Simplify the assessment of the competence of organisation.

The Approved Code of Practice (ACoP) has special legal status and gives practical advice for all those involved in construction work. If you follow the advice in the ACoP you will be doing enough to comply with the law in respect of those specific matters on which it gives advice.

The ACoP which includes a copy of the CDM regulations explains:

- The legal duties placed on clients, CDM co-ordinators, designers, principal contractors, contractors, self-employed and workers.
- The circumstances in which domestic clients do not have duties under CDM 2007 (but that the regulations still apply to those doing work for them).
- Gives information on the new role of CDM co-ordinator–a key project adviser for clients and responsible for coordinating the arrangements for health and safety during the planning phase of larger and more complex projects.
- Which construction projects need to be notified to HSE before work starts and gives information on how this should be done.
- How to assess the competence of organisations and individuals involved in construction work.
- How to improve co-operation and co-ordination between all those involved in the construction project and with the workforce.

- What essential information needs to be recorded in construction health and safety plans and files, as well as what shouldn't be included.

RATIONALE FOR THE RESEARCH

The subject of this dissertation developed from a personal interest in the Health and Safety aspect of the construction industry and the significant research within the industry on how to minimise the injuries and fatalities on site. Most of this research has concentrated on the best ways to improve the implementation process of the CDM regulations 2007, whether the changes are necessary from the CDM regulations 1994 and in terms of whether these changes have brought about the desired results and effect on the industry that it intended *I.e.* have the CDM regulations brought about an improvement in the safety of site work with regards to a clearer, faster and easier process of implementation.

Having looked at a wide range of literature and internet sites on the CDM regulations it is clear to me that there was evidence that the introduction of the CDM regulations was regarded by professionals as advantageous to the industry, however there are certain people that believe it has been less effective than it should have been on the industry. Although people have highlighted problems with the CDM regulations it is definitely an introduction by the government that is here to stay. This consideration has led me to undertake research into whether the new legislation has been implemented successfully and whether the changes are necessary. This was implemented under the research title; *the implementation of the CDM regulations under the CDM co-ordinators.*

The CDM regulations have the potential to make a significant impact on the industry with regards to health and safety however it is unrealistic to think that this can be achieved very quickly without encountering any problems in an area which has had much debate regarding this topic. It will take time before the new legislation will be free flowing however improvements in both effectiveness and efficiency can be expected as lessons are learnt. *E.g.* more home CDM co-ordinators becoming qualified.

Following its initial introduction in 1994, the Construction Design and Management Regulations (commonly known as the CDM Regulations) were re-introduced in April 2007, the revised Regulations are intended to make it easier for those involved in construction projects to comply with their health and safety duties.

The CDM Regulations are aimed at improving the overall management and co-ordination of health, safety and welfare throughout all stages of a construction project to reduce the large number of serious and fatal accidents and cases of ill health which happen every year in the construction industry. The HSE says that the new regulations emphasise planning and management to secure a safe project, rather than paperwork.

The Regulations place duties on all those who can contribute to the health and safety of a construction project. Duties are placed upon clients, designers and contractors with more power given to the CDM Coordinator in what is considered a more authorative and policing role.

The new regulations combine the Construction (Health, Safety and Welfare) Regulations and CDM into one single set of regulations. However, they also introduce some important changes to the safety regime.

These include:

- A new duty on designers to eliminate hazards and reduce risks, as far as is reasonably practicable. They will also have a duty to ensure that any workplace they design complies with the Workplace Regulations;
- A Client will no longer be able to appoint an agent to take on their legal duties and criminal liabilities, thereby making the CDM Coordinator role more advisory in helping to fulfill their duties to comply with the Regulations
- When Principal Contractors appoint contractors, they will have to tell those contractors how much time they have to prepare for on-site work;

- Contractors will have a similar duty towards those they appoint to work on-site, as well as being obligated to plan and manage their own work; and
- The role of Planning Supervisor, which carried responsibility for co-ordinating health and safety aspects of the design and the planning phase of the construction project, is to be replaced by a CDM Project Co-ordinator.

A CDM Co-ordinator, like a Planning Supervisor, has to be appointed by the client if a project lasts more than 30 days or involves more than 500 person days of work. But, unlike the Planning Supervisor, is required to advise and assist the client on how to fulfill their duties, especially on whether other duty-holders' arrangements are adequate. At the most, only the initial design work for the job should have been completed before the position is filled.

WHAT CDM APPLIES TO

The CDM 2007 Regulations apply to most common building, civil engineering and engineering construction work.

Construction work means the carrying out of any building, civil engineering or engineering construction work and includes:

- The construction, alteration, conversion, fitting out, commissioning, renovation, repair, upkeep, redecoration or other maintenance (including cleaning which involves the use of water or an abrasive at high pressure or the use of corrosive or toxic substances), de-commissioning, demolition or dismantling of a structure;
- The preparation for an intended structure, including site clearance, exploration, investigation (but not site survey) and excavation, and the clearance or preparation of the site or structure for use or occupation at its conclusion;
- The assembly on site of prefabricated elements to form a structure or the disassembly on site of prefabricated elements which, immediately before such disassembly, formed a structure;

- The removal of a structure or of any product or waste resulting from demolition or dismantling of a structure or from disassembly of prefabricated elements which immediately before such disassembly formed such a structure; and
- The installation, commissioning, maintenance, repair or removal of mechanical, electrical, gas, compressed air, hydraulic, telecommunications, computer or similar services which are normally fixed within or to a structure.

WHAT CDM DOES NOT APPLY TO

- Putting up and taking down marquees and similar tents designed to be re-erected at various locations.
- General maintenance of fixed plant, except when this is done as part of other construction work, or it involves substantial dismantling or alteration of fixed plant which is large enough to be a structure in its own right, for, example structural alteration of a large silo; complex chemical plant; power station generator or large boiler.
- Tree planting and general horticultural work.
- Positioning and removal of lightweight movable partitions, such as those used to divide open-plan offices or to create exhibition stands and displays.
- Surveying-this includes taking levels, making measurements and examining a structure for faults.
- Work to or on vessels such as ships and mobile offshore installations.
- Off-site manufacture of items for later use in construction work (for example roof trusses, pre-cast concrete panels, bathroom pods and similar prefabricated elements and components).
- fabrication of elements which will form parts of offshore installations.
- The construction of fixed offshore oil and gas installations at the place where they will be used.

WHAT EXTRA RESPONSIBILITIES DO THE CDM 2007 REGULATIONS PLACE ON CLIENTS?

While there are no new client duties; pre-existing ones have been strengthened, the HSE says:

Clients already had duties under HSWA 1974 and the Management of Health and Safety at Work Regulations 1999 to ensure construction projects were carried out safely.

However under the new CDM 2007, clients are explicitly instructed to take reasonable steps to ensure that:

- Construction risk can be carried out without risk to health and safety;
- Welfare arrangements are in place before work begins;
- Any structure designed for use as a workplace complies with the Workplace Regulations;
- Sufficient time and resources are allocated to achieve these duties and
- To indicate to contractors and designers how much time is available for planning and preparation before work starts.

An accompanying 'Approved Code of Practice' ACoP, due to be published some time from February 2007, gives simple advice on how to fulfil these duties, says the HSE. However, a client will no longer be able to appoint a 'client's agent' to take on their legal duties and criminal liabilities under CDM.

WHAT OTHER SIGNIFICANT CHANGES DO THE CDM REGULATIONS OF 2007 BRING IN?

The Executive believes that this provision of CDM 1994 was confusing. Even if the client appointed an agent, they continued to have duties and criminal liabilities under the HSWA 1974 and the MHSWR 1999. A client can still appoint a professional to carry out their duties but the legal responsibility to comply with CDM stays with the client.

Other changes to the regulatory regime include:

Duty holders must be sure that anyone they appoint to carry out or manage design or construction is competent; and Duty holders themselves need to ensure that they are competent.

HOW CAN I ENSURE MY BUSINESS IS CDM COMPLIANT?

To ensure that your company meets the CDM regulations, is recommended to have a CDM audit carried out by a third party. A CDM compliance audit provides an objective third party view of your company strengths and weaknesses in this area. The CDM compliance audit takes part in two separate stages. The initial approach is to gather information; this is followed by a detailed evaluation which will be presented in a formal report. If changes and alterations are noted in relation to CDM regulations, an objective project management company can assist with new implementation measures

WILL CDM HELP WITH DEMOLITION?

CDM regulations play an invaluable role in the life cycle of development and demolition that occurs in process plants and manufacturing sites. The safe demolition of disused facilities is crucial and the CDM regulations provide a structured base from which to work.

Many companies are understandably nervous of undertaking the demolition work that may be required at a site, especially when they consider the CDM regulations they have to adhere to. There is often a belief that costs will be prohibitive and the risks difficult to manage. In this instance, an impartial project management company can offer an important support service in this situation and provide services to assist with the adherence of CDM regulations:

- Fulfilling obligations under CDM
- Undertaking a full scoping and pricing study
- Selecting competent demolition contractors
- Managing the demolition project in reference to CDM
- Assessing the environmental impact

CDM TRAINING FOR BUSINESS

As well as understanding the ways in which CDM regulations affect your business, there is also the need for a constant monitoring or to ensure a safety-driven business culture. Training staff to be aware of and monitor CDM regulations can enable CDM safety to become part of everyday life.

A project management/training company can support your CDM learning programme by a number of different techniques:

- Tool box talks for staff
- CDM training for Duty Holders
- Alliance and team building events to spread the safety message
- CDM mentoring for inexperienced construction staff

WHO CAN HELP ME ENSURE I COMPLY AND IMPROVE IN LINE WITH CDM REGULATIONS?

Since the inception of the 1994 CDM regulations and the new regulations which came in force in April 2007, PROjEN have supported many clients to ensure compliance with the legislation. PROjEN's involvement with clients is flexible and multi-faceted. We can provide CDM support as a stand alone role, such as CDM Coordinator on internal projects. Alternatively the role of principal contractor, CDM Coordinator and designer can be represented within a turnkey project or alliance scenario. In addition to working alongside our clients, PROjEN have been active in the interpretation and development of the legislation via our role as a member of the CDM Duty Holders Support Group and our involvement in many safety initiatives. This means that our advice on procedural and documentation issues is comprehensive and well developed.

This covers areas such as:

- Support on the development of Risk assessment
- Development of Health and Safety Plans and Files
- Compliance and reporting
- Auditing of the Construction Site
- Chairing and Supporting the Design Risk Assessments Process

SPECIALIST CONSULTANTS IN CDM SUPPORT AND TRAINING

PROjEN PLC are a well established project management

company recognised as being leaders in the delivery of industrial projects on a stand alone/turnkey basis or working alongside client companies.

As part of our service offering, PROjEN deliver a number of Business Improvement Products and Services which have been carefully selected to add real benefits. These products/ services have their foundation in the successful implementation of projects over the last three decades, a statement given credence through the recognition of the 2005 Bentley BE Award for Excellence of New Technology Adoption, ECI Contractor of the Year in 2003, ECI ACTIVE Project of the Year Awards in 2004, 2006 and 2008 as well as another 'Royal Society for the Prevention of Accidents' (RoSPA) Gold Medal in 2009–the eighth consecutive Gold Award the Company has received. PROjEN have also been awarded Vale Royal Business of the year 2006, the High Sheriff of Cheshire award for Enterprise in 2007 and received Carbon Trust Accreditation in 2007.

PROjEN can work independently or alongside the client project owners, providing expert advice from day one. We supply crucial resource when it is needed, control budgets and programmes and ultimately deliver the project as the client originally intended. Our core business is the successful delivery of projects from feasibility, capital justification, front-end engineering, detailed design, through to full turnkey project completion, providing the client with single point responsibility.

In an age of specialisation, PROjEN can also offer a full design responsibility across all disciplines in addition to Construction Supervision, Health and Safety, Procurement and Commissioning. Our extensive experience across a wide range of industries encourages cross fertilisation of ideas and adoption of best practice techniques. Our up to date knowledge and translation of current legislation and CDM regulations which allows PROjEN to provide practical help and advice at costs sufficient to satisfy current regulations.

BUILDING REGULATIONS BRITISH STANDARDS

The histories of building regulations differ all over the

world. In Britain, for instance, the Great Fire of 1666 saw the introduction of regulations designed to prevent a repetition of that disastrous event by requiring an early form of compartmentation together with stone construction. In Qatar the requirement for building regulations or standards were tied to the development of the State with the growth and influence of the merchants having a notable affect on their character and scope. The early years of development, essentially, post-1945, saw a burgeoning of commercial interests with the registration of 'General Merchants and Building Contractors', the legal title reflecting the inter-relationship of materials and the rapidly evolving building industry. Commonly, those who brought materials into the country sought to use them in the new buildings required for the growing population.

Many of the materials introduced reflected the interests and contacts of local entrepreneurs. The larger organisations associated with the oil industry imposed standards in terms of specifications, and this in turn linked many of the materials to sources in the United Kingdom where there were both accepted material standards as well as codes of practice and building regulations. The consultants and contractors who began to construct the larger developments used these as they were the regulations and standards familiar to them and to which they had to adhere for professional purposes, many of the projects being controlled by reference to Bills of Quantities.

POOR PRACTICES AND MATERIALS

More notes have been written on this subject on the page dealing with the building industry and its history in Qatar. Generally the problem had three components: specification, the methods of construction, and supervision. Much of this had to do with a basic ignorance of sound building techniques this, in turn, being a result of obtaining labour and supervision themselves ignorant in varying degrees as well as a lack of understanding by clients, many of them their own contractors.

But there was also an issue with the importation and re-use of materials unfit for the various jobs. Again, partly this was due to a lack of understanding or experience of the suppliers, but it was also a product of the dumping of materials by less than scrupulous middlemen outside the country. At the back of all this was the desire to make a profit from what was understood locally as a burgeoning market.

The result of the initial building boom was, as you might expect, buildings that responded badly to the environmental conditions, that represented poor value for money and that did not create a good impression for the developing country.

THE BEGINNING OF STANDARDS

The new government structures of the latter half of the twentieth century introduced professional staff from abroad. The Ministry of Public Works and the Ministry of Electricity and Water had a significant impact in improving standards of construction for works both above and below ground, though they were mainly concerned with government projects. The Ministry of Municipal Affairs also had an interest in the control of building, but its standards did not keep in step with those of the previous two ministries due mainly to a lack of suitably qualified staff.

Many of the original professional staff were British and brought with them to Qatar an inherent knowledge and understanding of British standards, codes of practice and building regulations. Much of the initial infrastructural development was, therefore, installed to British Standards and specifications. This continued with many of the Government's buildings being constructed with these standards though some new buildings that were designed by architects other than British, brought with them their own standards and specifications. As a general principle there was no problem with this approach as the standards of design and specification were coherent and covered the concerns that usually govern design and construction. The reason for this approach was simply that the State had not yet assembled a coherent set of regulations or codes, and did not yet have the staff to apply them; there was a team of inspectors at the Ministry of Municipal Affairs, but they only dealt with small scale projects that were mostly private. So, in the early stages of development there were a range of standards being applied to public and private buildings. At the same time there was no education of the contractors other than in response to market forces that–despite the ease with which property was being rented–saw an increasing division between up-market and down-market accommodation. Having said that, there are still complaints in the expatriate press relating to the standards of finish by those renting up-market accommodation.

QATAR BUILDING REGULATIONS

As mentioned on the page relating to planning in Qatar, building development in Qatar began to increase in pace after the instability within the region created by the approach to, and the consequent Second World War. Most of the development in the peninsula was centred on Doha so the nascent government created the Municipality of Qatar under Law no. 11 of 1963 with responsibility for control within four defined areas. Later in the year this title was amended under Law no. 15 to create the Municipality of Doha.

Its remit covered the:

- Building process,
- Gardens,
- Public health, and
- Accounts.

This was the first stage in the production of requirements relating to the construction of buildings. It specifically required the submission of site location plans with the intent of controlling buildings through licensing, though at that time there was no effective process for enforcing either any regulations or errors in setting out building sites other than through the courts. This state of affairs continued until Qatar declared its independence in 1971 when a ministry structure was introduced, and the Ministry of Municipal Affairs was established having responsibility for all municipalities in the peninsula, and the process of tightening up licensing began. It should be borne in mind that, at the time of the 1970 Census, there were only six conurbations with over a thousand head of population–Doha, New Rayyan, Old Rayyan, al-Gharafa, al-Khor and al-Wakra with Doha by far the largest. With New Rayyan, Old Rayyan and al-Gharafa being, in effect, suburbs of Doha and, consequently, within a greater Doha conurbation, outside Doha only al-Wakra and al-Khor had populations greater than a thousand:

Conurbation	Population according to the 1970 Census
Doha	83,300
Old Rayyan	2,900
New Rayyan	2,900
al-Gharafa	1,100
al-Khor	1,900
al-Wakra	1,800

1972 saw the introduction of planning consultants with a remit to establish planning standards for Doha and Umm Said specifically, and the rest of the peninsula strategically, but this did not include the production of building regulations.

The United Nations report of 1997 on the implementation of Agenda 21–a 300 page plan for achieving sustainable development in the 21st century–pointed up the lack of a coherent approach to the regulation of construction in Qatar. This they established based on a report submitted to them from the Ministry of Municipal Affairs and Agriculture's Environment Department. They noted:

Decision-Making Structure: New constructions are required to take a building permit which has to be approved by the concerned Municipality, Planning Department and other service authorities including Electricity, Water and Civil Defence. A completion certificate is also required to ensure compliance with all regulations. Legislation on a number of issues has already been passed.

The main ones are as follows:

- Law No. 4–1985 and its amendments, related to controlling buildings.

- Ministerial Order No. 2–1989, related to precautions to be taken for public safety to avoid hazards from building construction.
- Ministerial Order No. 7–1989, related to technical and architectural specifications for buildings.
- Law No 3-1975 and its amendments, related to commercial, industrial and public buildings.
- Law No 8-1974 and its amendments, related to general cleanliness.
- Ministerial Order No 2-1989, related to the transport of debris, solid and liquid waste.
- In addition to the above legislation, the Planning Department of the Ministry of Municipal Affairs and Agriculture has published several planning and building regulations under the existing law giving specifications for the main development types in urban areas. These include:
- Regulations for Flats and Flat Complexes (Draft)
- Regulations for Villa and Villa Complexes (Draft)
- Qatar Commercial Development Manual (Draft)
- Planning Regulations for Residential Developments in Al Rayyan Municipality
- Planning Regulations for Commercial Developments in Al Rayyan Municipality
- Sub-division Regulations
- Interim Zoning Regulation
- The main constraints facing implementation of legislation and criteria regulating construction in urban areas can be summarized as follows:
- The existing legislation is not comprehensive. There are still several areas for which legislation need to be enacted.
- Lack of proper mechanisms for enforcement of existing legislation which includes lack of experienced personnel and funding.
- Many of the regulatory documents do not have legal status.

It is apparent that the Government is in the process of ensuring that a more coherent and effective system of controls for regulating materials and building is being developed and staffed, and it is understood that a degree of equivalency will be accepted from reputable foreign authorities in order to optimise the amount of testing carried out in Qatar.

CONSTRUCTION PERMISSIONS

In the meantime, and in response to Law no. 16 of 2002, Qatar General Organization for Standards and Metrology was established with a remit to produce a set of Qatari Standards for a number of items, including building materials. This is now being implemented. In order to demonstrate how the process works, here is a schedule of the activities, timing and costs involved in the process of building a small warehouse in January 2009, reproduced with the permission of the copyright holders, the World Bank's International Bank of Reconstruction and Development:

STANDARD

The British Standards Institution (BSI) is the UK National Body providing the organisation, facilities and structure for the preparation of UK National Standards. British Standards Online is the most authoritative current site for all BSI publications.

Health and Safety staff can readily gain information regarding British Standards, by accessing the HSE Intranet site, going into HSE Information Services.

British Standards Online has three levels of user access:

- *Registered Guest*: In addition to searching and viewing results, registering for Free also allows you to view summaries and order hardcopy documents online.
- *Subscriber*: Become a Subscriber and get instant unlimited online access to standards. For all HSE staff, information from British Standards is available without having to subscribe.
- In addition to standards the site also includes:
- Technical handbooks

- Codes of practice
- Guidelines
- Specifications for products, dimensions, and performance
- Glossaries

Please note the following definitions as applied to British Standards:

- *Withdrawn*: The standard is no longer current
- *Obsolescent:* The standard is not recommended for use for new equipment, but is retained to provide for the servicing of existing equipment that is expected to have a long working life.

BS Series		**Title**
BS 6949:1991	Confirmed, current (01/10/2008) Replaces	Specification for bitumen-based coatings orcold application excluding use in contact with potable water
BS 470-1984	(Current)	Specification for inspection, access and entry openings for pressure vessels
BS 476-3-1975	(Current)	Fire tests on building material and structures. External fire exposure roof test
BS 476-4-1989	(Current)	Fire tests on building materials and structures. Non-combustibility test for materials
BS 476-6-1989	(Current)	Fire tests on building materials and structures. Method of test for fire propagation for products
BS 476-7-1997	(Current)	Fire tests on building materials and structures. Method of test to determine the classification of the surface spread of flame of products
BS 476-10-1983	(Current)	Fire tests on building materials and structures. Guide to the principles and application of fire testing
BS 476-11-1982	(Current)	Fire tests on building materials and structures. Method for assessing the heat

		emission from building materials
BS 476-12-1991	(Current)	Fire tests on building materials and structures. Method of test for ignitability of products by direct flame impingement
BS 476-13-1987	(Current)	Fire tests on building materials and structures. Method of measuring the ignitability of products subjected to thermal irradiance
BS 476-15-1993	(Current)	Fire tests on building materials and structures. Method for measuring the rate of heat release of products
BS 476-20-1987	(Current)	Fire tests on building materials and structures. Method for determination of the fire resistance of elements of construction (general principles)
BS 476-21-1987 elements	(Current)	Fire tests on building materials and structures. Method for determination of the fire resistance of loadbearing of construction
BS 476-22-1987	(Current)	Fire tests on building materials and structures. Methods for determination of the fire resistance of non-loadbearing elements of construction
BS 476-23-1987	(Current)	Fire tests on building materials and structures. Methods for the determination of the contribution of components to the fire resistance of a structure
BS 476-24-1987	(Current)	Fire tests on building materials and structures. Method for the determination of the fire resistance of ventilation ducts
BS 476-31.1-1983	(Current)	Fire tests on building materials and structures. Methods for measuring smoke penetration through doorsets and shutter assemblies. Method of measurement under ambient temperature conditions
BS 476-32-1989	(Current)	Fire tests on building materials and

		structures. Guide to full scale fire tests within buildings
BS 476-33-1993	(Current)	Fire tests on building materials and structures. Full-scale room test for surface products
BS 2594-1975	(Current, superseded)	Specification for carbon steel welded horizontal cylindrical storage tanks
BS 2654-1989	(Current) (formerly BS 2654-1984)	Specification for manufacture of vertical steel welded non-refrigerated storage tanks with butt-welded shells for the petroleum industry
BS 2633-1987	(Current)	Specification for class 1 arc welding of ferritic steel pipework for carrying fluids
BS 2971-1991	(Current)	Specification for class 2 arc welding of carbon steel pipework for carrying fluids
BS 3293-1960	(Current)	Specification for carbon steel flanges (over 24 inches nominal size) for the petroleum industry
BS 3416-1991	Confirmed, current (01/10/2008) Replaces BS 3416:1988	Specification for bitumen-based coatings for cold application, suitable for use in contact with portable water
BS 3492-1987	(Current)	Specification for road and rail tanker hoses and hose assemblies for petroleum products, including aviation fuels
BS 3693: 1992	(Current)	Recommendations for designs of scales and indexes on analogue indicating instruments
BS 4250-1997	(Current)	Specification for commercial butane and commercial propane
BS 4871-3-1985	(Current)	Specification for approval testing of welders working to approved welding procedures. Arc welding of tube to tube plate joints in metallic materials

BS 4872-1-1982	(Current)	Specification for approval testing of welders when welding procedure approval is not required. Fusion welding of steel
BS 4872-2-1976	(Current)	Specification for approval testing of welders when welding procedure approval is not required. TIG or MIG welding of aluminium and its alloys
BS 4994-1987	(Current, superseded) (replaced by BS EN 13923:2005, BS EN 13121-3:2008, Replaced by BS EN 13923:2005 and BS EN 13121-3:2008 but remains current.)	Specification for design and construction of vessels and tanks in reinforced plastics
BS 5306-0-1986	(Current) partially replaced by BS EN 671-3-2000)	Fire extinguishing installations and equipment on premises. Guide for the selection of installed systems and other fire equipment
BS 5306-1-1976	(Current, partially replaced by BS EN 671-3 - 2000)	Fire extinguishing installations and equipment on premises. Hydrant systems, hose reels and foam inlets
BS 5306-2-1990	(Current)	Fire Extinguishing installations and equipment on premises. Specification for sprinkler systems
BS 5306-3-2000	(Current)	Fire extinguishing installations and

		equipment on premises. Maintenance of portable fire extinguishers. Code of practice
BS 5306-4-2001	(Current)	Fire extinguishing installations and equipment on premises. Specification for carbon dioxide systems
BS 5306-5.1-1992	(Current)	Code of practice for fire extinguishing installations and equipment on premises. Halon Systems. Specification for Halon 1301 total flooding systems
BS 5306-5.2-1984	(Current)	Code of practice for fire extinguishing installations and equipment on premises. Halon Systems. Halon 1211 total flooding systems
BS 5306-6.1-1988	(Current)	Fire extinguishing installations and equipment on premises. Foam systems. Specification for low expansion foam systems
BS 5306-6.2-1989	(Current)	Fire extinguishing installations and equipment on premises. Foam systems. Specification for medium and high expansion foam systems.
BS 5306-7-1988	(Current, superseded)	Fire extinguishing installations and equipment on premises. Specification for powder systems
BS 5306-8-2000	(Current)	Fire extinguishing installations and equipment on premises. Selection and installation of portable fire extinguishers. Code of practice
BS 5351: 1986	(Current)	Specification for steel ball valves for the petroleum, petrochemical and allied industries
BS 5415-1-1985	(Current)	Safety of electrical motor-operated industrial and commercial cleaning appliances. Specification for general requirements

BS 5493-1977	(Current)	Code of practice for protective coating of iron and steel structures against corrosion
BS 5501-8:-1988	(Current)	Electrical apparatus for potentially explosive atmospheres. Encapsulation 'm'
BS 5908-1990	(Current, obsolescent)	Code of practice for fire precautions in the chemical and allied industries
BS 5958-1-1991 BS 5958-2-1991	(Current, superseded) (Current, superseded)	Code of practice for control of undesirable static electricity. Code of practice for control of undesirable static electricity. Recommendations for particular industrial situations
BS 6349-1:-2000	(Current)	Maritime structures. Code of practice for general criteria
BS 6349 -2:-1988	(Current)	Maritime structures. Design of quay walls, jetties and dolphins
BS 6349 -3:-1988	(Current)	Maritime structures. Design of dry docks, locks, slipways and shipbuilding berths, shiplifts and dock and lock gates
BS 6349 -4:-1994	(Current)	Maritime structures. Code of practice for design of fendering and mooring systems
BS 6349 -5:-1991	(Current)	Marine structures. Code of practice for dredging and land reclamation
BS 6349 -6:-1989	(Current)	Maritime structures. Design of inshore moorings and floating structures
BS 6349 -7:-1991	(Current)	Maritime structures. Guide to the design and construction of breakwaters
BS 6399-1-1996	(Current)	Loading for buildings. Code of practice for dead and imposed loads
BS 6399-2-1997	(Current)	Loading for buildings. Code of practice for wind loads
BS 6399-3-1988	(Current)	Loading for buildings. Code of practice for imposed roof loads
BS 6464-1984	(Current)	Specification for reinforced plastics pipes, fittings and joints for process plants

BS 6467-1-1985	(Current)	Electrical apparatus with protection by enclosure for use in the presence of combustible dusts. Specification for apparatus
BS 6467-2-1988	(Current)	Electrical apparatus with protection by enclosure for use in the presence of combustible dusts. Guide to selection, installation and maintenance
BS 6739: 1986	(Current)	Code of practice for instrumentation in process control systems: installation design and practice
BS 6651-1999	(Current, work in hand)	Code of Practice for protection of structures against lighting
BS 6656-2002	(Current)	Assessment of inadvertent ignition of flammable atmospheres by radio frequency radiation. Guide
BS 6990-1989	(Current)	Code of practice for welding on steel pipes containing process fluids or their residuals
BS 7430-1998	(Current)	Code of Practice for earthing
BS 7445-1: 2003	(Current)	Description and measurement of environmental noise. Guide to quantities and procedures
BS 7445-2: 1991	(Current)	Description and measurement of environmental noise. Guide to the acquisition of data pertinent to land use
BS 7445-3: 1991	(Current)	Description and measurement of environmental noise. Guide to application to noise limits
BS 7535-1992	(Current)	Guide to the use of electrical apparatus complying with BS5501 or BS 6942 in the presence of combustible dusts
BS 7671-2001	(Current)	Requirements for electrical installations. IEE wiring regulations. Sixteenth edition

BS 7777-1- 1993	(Current)	Flat-bottomed, vertical, cylindrical storage tanks for low temperature service. Guide to the general provisions applying for design, construction, installation and operation
BS 7777-2-1993	(Current)	Flat-bottomed, vertical, cylindrical storage tanks for low temperature service. Specification for the design and construction of single, double and full containment metal tanks for the storage of liquefied gas at temperatures down to -1650C
BS 7777-3-1993	(Current)	Flat-bottomed, vertical, cylind.ical storage tanks for low temperature service. Recommendations for the design and construction of prestressed and reinforced concrete tanks and tank foundations, and for the design and installation of tank insulation, tank linings and coatings
BS 7777-4-1993	(Current)	Flat-bottomed, vertical, cylindrical storage tanks for low temperature service. Specification for the design and construction of single containment tanks for the storage of liquid oxygen, liquid nitrogen and liquid argon
BS EN Series		**Title**
BS EN ISO 15614-3:2008	(Current)	Specification and qualification of welding procedures for metallic materials. Welding procedure test. Fusion welding of non-alloyed and low-alloyed cast irons.
BS EN 287-1: 2004	(Current) (formerly BSEN 287-	Qualification test of welders. Fusion welding. Steels

	1: 1992)	
BS EN 288-1-1992	(Current)	Specification and approval of welding procedures for metallic materials. General rules for fusion welding
BS EN 288-6-1995	(Current)	Specification and approval of welding procedures for metallic materials. Approval related to previous experience
BS EN 288-8-1995	(Current)	Specification and approval of welding procedures for metallic materials. Approval by a pre-production welding test
BS EN 288-9-1999	(Current)	Specification and approval of welding procedures for metallic materials. Welding procedure test for pipeline welding on land and offshore site butt welding of transmission pipelines
BS EN 894-1: 1997	(Current)	Safety of machinery. Ergonomics requirements for the design of displays and control actuators. General principles for human interactions with displays and control actuators
BS EN 894-2: 1997	(Current)	Safety of machinery. Ergonomics requirements for the design of displays and control actuators. Displays
BS EN 954-1-1997	(Current)	Safety of machinery. Safety related parts of control systems. General principles for design
BS EN 1050-1997	(Current)	Safety of machinery. Principles for risk assessment
BS EN 1092-1: 2002	(Current)	Flanges and their joints. Circular flanges for pipes, valves, fittings and accessories, PN designated. Steel flanges
BS EN 1127-1-1998	(Current)	Explosives atmospheres. Explosion prevention and protection. Basic concepts and methodology

BS EN 1127-2-2002	(Current)	Explosive atmospheres. Explosion prevention and protection. Basic concepts and methodology for mining
BS EN 1539: 2000	(Current)	Dryers and ovens in which flammable substances are released. Safety requirements.
BS EN 1755: 2000	(Current)	Safety of industrial trucks. Operation in potentially explosive atmospheres. Use in flammable gas, vapour, mist and dust
BS EN 13463-	(Current) 1: 2001	Non-electrical equipment for potentially explosive atmospheres. Basic method and requirements
BS EN 13463-	(Current) 5: 2003	Non-electrical equipment for potentially explosive atmospheres. Protection by constructional safety 'c'
BS EN 13463-8: 2003	(Current)	Non-electrical equipment for potentially explosive atmospheres. Protection by liquid immersion 'k'
BS EN 13480-1-2002	(Current) (formerly BS 806-1993)	Metallic industrial piping. General
BS EN 13480-2-2002	(Current) (formerly BS 806-1993)	Metallic industrial piping. Materials
BS EN 13480-3-2002	(Current) (formerly BS 806-1993)	Metallic industrial piping. Design and calculation
BS EN 13480-4-2002	(Current) (formerly BS 806-1993)	Metallic industrial piping. Fabrication and installation
BS EN 13480-5-2002	(Current) (formerly BS 806-1993)	Metallic industrial piping. Inspection and testing
BS EN 50014-1998	(Current) (Formerly	Electrical apparatus for potentially explosive atmospheres. General

	BS 4683-1: 1971)	requirements
BS EN 50015-1998	(Current)	Electrical apparatus for potentially explosive atmospheres. Oil immersion "o"
BS EN 50016-2002	(Current, Superceded, Obsolescent)	Electrical apparatus for potentially explosive atmospheres. Pressurised apparatus "p"
	(Replaced by BS EN 60079-2: 2004, but remains current as obsolescent)	
BS EN 50017-1998	(Current)	Electrical apparatus for potentially explosive atmospheres. Powder filling "q"
BS EN 50018-2000	(Current)	Electrical apparatus for potentially explosive atmospheres. Flameproof enclosure 'd'
BS EN 50019-2000	(Current) (formerly BSEN 50019-1994)	Electrical apparatus for potentially explosive atmospheres. Increased safety 'e'
BS EN 50020-2002	(Current)	Electrical apparatus for potentially explosive atmospheres. Intrinsic safety 'i'
BS EN 50021-1999	(Current) (formerly BS 6941-1988) (Proposed for obsolescence)	Electrical apparatus for potentially explosive atmospheres. Type of protection 'n'
BS EN 50281-1-2-1999	(Current) (formerly BS 6467-2-1985)	Electrical apparatus for use in the presence of combustible dust. Electrical apparatus protected by enclosures. Selection, installation

		and maintenance
BS EN 50281-2-1: 1999	(Current)	Electrical apparatus for use in the presence of combustible dust. Test methods. Methods of determining minimum ignition temperatures
BS EN 50281-3: 2002	(Current)	Electrical apparatus for use in the presence of combustible dust. Classification of area where combustible dusts are or may be present
BS EN 60073: 2002	(Current)	Basic and safety principles for man machine interface, marking and identification. Coding principles for indicators and actuators
BS EN 60079-1-2004	(Current) (Replaces BS EN 50018-2000, which remains current)	Electrical apparatus for explosive gas atmospheres. Flameproof enclosures 'd'
BS EN 60079-2-2004	(Current)	Electrical apparatus for explosive gas atmospheres. Pressurised enclosures "p"
BS EN 60079-10-2003	(Current) (formerly BS EN 60079-10-1996)	Electrical apparatus for explosive gas atmospheres. Classification of hazardous areas
BS EN 60079-14-1997	(Current) (formerly BS 5345-1)	Electrical apparatus for explosive gas atmospheres. Electrical installations in hazardous areas (other than mines)
BS EN 60079-15: 2005	(Current) (This replaces	Electrical apparatus for explosive gas atmospheres. Construction, test and marketing of type protection 'n'

	BS 6941: 1988) Work in hand	electrical apparatus.
BS EN 60079-17-2003	(Current) (formerly BS EN 60079-10-1997)	Electrical Apparatus for explosive gas atmospheres. Inspection and maintenance of electrical installations in hazardous areas (other than Mines)
BS EN 60079-18-2004	(Current) (Replaces BS 5501-8-1988, which remains current)	Electrical apparatus for explosive gas atmospheres. Construction, test and marking of type of protection encapsulation 'm' electrical apparatus
BS EN 60529-1992	(Current)	Specification for degrees of protection provided by enclosures (IP code)
BS EN 61000 -4-1:2001	(Current)	Electromagnetic compatibility (EMC). Testing and measurement techniques. Overview of IEC 61000-4 series
BS EN 61131 -2:2003	(Current) (Replaces- BS EN 61131-2: 1995)	Programmable controllers. Equipment requirements and tests
BS EN 61131-3 -2003	(Current) (formerly BS EN 61131-3-1993)	Programmable controllers. Programming languages
BS EN 61131-5 -2001	(Current)	Programmable controllers. Communications
BS EN 61131-7 -2001	(Current)	Programmable controllers. Fuzzy control programming

BS EN 61241-14: 2004	(Current) (Replaces 50281-1-2: 1999 but this still current)	Electrical apparatus for use in the presence of combustible dust. Selection and Installation.
BS EN 61508-1-2002	(Current) (Identical IEC 61508-1998)	Functional safety of electrical/ toelectronic/programmable electronic safety-related systems. General requirements
BS EN 61508-2-2002	(Current) (Identical IEC 61508-2-2000)	Functional safety of electrical/ to electronic/programmable electronic safety-related systems. Requirements for electrical/electronic/programmable electronic safety-related systems
BS EN 61508-3-2002	(Current) (Identical IEC 61508-3-1998)	Functional safety of electrical/ toelectronic/programmable electronic safety-related systems. Software requirements
BS EN 61508-4-2002	(Current) (Identical IEC 61508-4-1998)	Functional safety of electrical/ toelectronic/programmable electronic safety-related systems. Definitions and abbreviations
BS EN 61508-5-2002	(Current) (Identical IEC 61508-5-1998)	Functional safety of electrical/ toelectronic/programmable electronic safety-related systems. Examples of methods for the determination of safety integrity levels
BS EN 61508-6-2002	(Current) (Identical IEC 61508-6-1998)	Functional safety of electrical/ to electronic/programmable electronic safety-related systems. Guidelines on the application of IEC 61508-2 and 61508-3
BS EN 61508-7-2002	(Current) (Identical IEC 61508-	Functional safety of electrical/ to electronic/programmable Electronic safety-related systems. Overview of

	7-2000)	techniques and measures
BS EN ISO Series		**Title**
BS EN ISO 4126-2003	(Current) (formerly BS 2915-1990	Safety devices for protection against excessive pressure. Bursting disc safety devices
BS EN ISO 4126-1: 2004	(Current) (Replaces BS 6759-1: 1984 BS 6759-2: 1984 BS 6759-3: 1984)	Safety devices for protection against excessive pressure. Safety valves
BS EN ISO 9606-2: 2004	(Current) (Replaces BS EN 287-2: 1992)	Approval testing of welders. Fusion welding. Aluminium and aluminium alloys.
BS EN ISO 10497: 2004	(Current) Replaces BS 6755-2: 1987	Testing of Valves. Fire type-testing requirements
BS EN ISO 15610-2003	(Current) (formerly BS EN 288-5: 1995)	Specification and qualification of welding procedures for metallic materials. Qualification based on tested welded consumables.
BS EN ISO 15612-2004	(Current) (Replaces BS EN 228-7: 1995)	Specification and qualification of welding procedures for metallic materials. Qualification by adoption of a standard welding procedure.
BS EN ISO 15609-1:2004	(Current) Replaces BS EN 288-2: 1992	Specification and qualification of welding procedures for metallic materials. Welding procedure specification. Arc welding
BS EN ISO	(Current)	Specification and qualification

15614-1:2004	(Replaces-BS EN 288-3: 1992)	of welding procedures for metallic materials. Welding procedure test. Arc and gas welding of steels and arc welding of nickel and nickel alloys
BS EN ISO 15614-2:2005	(Current) (Replaces-BS EN 288-4: 1992)	Specification and qualification of welding procedures for metallic materials. Welding procedure test. Arc welding of aluminium and its alloys.
BS EN ISO/IEC 17020: 2004	(Current) Replaces BS EN 45004: 1995	General criteria for the operation of various types of bodies performing inspection
BS IEC Series		**Title**
BS IEC 60079-19:1993, IEC 60079-19: 1993	(Current)	Electrical apparatus for explosive gas atmospheres. Repair and overhaul for apparatus used in explosive atmospheres (other than in mines or explosives)
PD SERIES		**Title**
PD 5500-2003	(Current) (formerly BS 5500-2000)	Specification for unfired fusion welded pressure vessels
PS IEC TR 61000-1-3:2002	(Current)	Electromagnetic compatibility (EMC). General. The effects of high-altitude EMP on civil equipment and systems

EUROPEAN STANDARDS

REGULATIONS

Regulation is usually introduced when it is recognised that market failures would not allow economic instruments alone to reach the objective of the energy or environmental policy. In general, regulations impose minimum efficiency standards by law and/or governmental decree, or introduce energy efficient practices (technical and behavioural/managerial), as well as providing systematic information to consumers (*e.g.* energy audits, labels).

Regulations can be set at the national level, at the level of a group of countries, or at the level of a sub-national region inside a federal country. There are also other regulations which are not specifically targeted at energy efficiency, but which can nonetheless influence (*e.g.* speed limits, maximum weight of trucks).

REGULATIONS FOR BUILDINGS

Most European countries have set up mandatory energy efficiency standards for new dwellings and service sector buildings. A new Directive on the energy performance of buildings introduce now in all EU countries harmonised standards for new buildings and make mandatory buildings certificates for the sale or rent of dwellings.

In half of the other OECD countries in Asia and America, there are mandatory and in the other half voluntary standards. Some non-OECD countries outside Europe have recently established mandatory or voluntary standards for service buildings: Singapore and the Philippines were among the first, followed by Algeria, Malaysia, Egypt and Syria for instance. In most countries, standards exist for both dwellings and service sector buildings, except in Africa and in Asia where most often standards only apply to non-residential buildings.

The situation in these two regions is explained by the fact that commercial buildings account for the largest share of energy consumption. Altogether, about 50% of the countries surveyed had mandatory or voluntary standards for new non-residential buildings.

Thermal building codes have been changing over time from simple standards on building components to more complex standards, including for the most advanced countries energy performance standard. These performance standards consider the whole building as a system and also include building equipment such as heating and air conditioning systems, ventilation, water heaters, and in some countries even pumps and elevators (maximum energy consumption per m3 or m2/ year). Most building codes now are performance based (*e.g.* present standards in California, Germany and France, or the EU building Directive). These types of standards can be implemented jointly with standards on specific equipment or materials (insulation, windows, boilers), in order to ensure the use of the most efficient equipment in the retrofitting of existing buildings (*e.g.* France).

Revisions in thermal building codes have become increasingly regular. For instance, over the past 30 years, standards have been reinforced three or four times in most EU-15 countries, including some very recent revisions, and independent from the oil price level. The effort is not finished yet, as most EU countries have improved their standards since the year 2000. In addition, the new EU building directive has for the first time provided for a mandatory revision every five years.

The cumulative energy savings achieved for new dwellings, compared to dwellings built before the first oil shock, is about 60% on average in the EU. The additional savings that are targeted with future revisions in the standards are also impressive, at 20-30%.

Relatively few countries have carried out evaluations of their building codes. According to the few studies available, it seems that the actual energy performance of new buildings is below what could be expected from the building regulations. This can be explained by behavioural factors (such as higher indoor temperatures, more rooms heated, or longer heating period over the year) and by a non-compliance with the building regulation.

Only a few countries have estimated the additional costs that each round of new building codes has caused. Nevertheless, from the few results available, the additional costs are limited to a few percentage points, if any at all. Measures on buildings focused so far on new buildings. As new buildings represent a small share of the existing stock, building standards can only have a slow impact on the short term, which however becomes significant in the long-term. A more recent trend is to extend regulations to existing buildings and impose the introduction of energy efficiency certificates for existing buildings; each time there is a change of tenant or a sale. Such a measure was introduced in Denmark some years ago and extended recently to all EU countries with the Directive on Buildings. These certificates enable the buyer to obtain information about the energy consumption of the dwelling they are going to buy or rent.

LABELLING AND EFFICIENCY STANDARDS FOR HOUSEHOLD ELECTRICAL APPLIANCES

To slow down or even reverse the trend in the electricity consumption of households, many countries have introduced labelling programmes and minimum energy performance standards for a selection of electrical appliances. Most countries first focused on refrigerators, along with air conditioners in certain countries, since they account for a large part of the household electricity consumption (in Europe, 20 30% depending on the country). Now these measures cover a greater number of equipment: lighting, washing machines, dryers, dishwashers and water heaters.

Labelling programmes are designed to provide consumers with information, which enables them to compare the energy efficiency of the different appliances on sale. They aim at modifying the selection criteria of consumers by drawing their attention to the energy consumption of household appliances. Labelling programmes however cannot sufficiently transform the market and are usually completed by minimum performance standards in the great majority of countries.

The aim of performance standards is to improve the energy efficiency of new appliances either by imposing a minimum energy efficiency rating to remove the least efficient products from the market, Minimum Energy Performance Standards (MEPS), or by requiring sales-weighted average energy efficiency improvements as "target values" *e.g.* "Top Runner Programme" in Japan. Target values are more flexible as they allow the sale of less efficient equipment provided other models with a higher efficiency rating are also offered for sale.

As an alternative to the regulatory process, make use of agreements with appliance manufacturers (voluntary or negotiated), which also improve the energy efficiency of appliances (*e.g.* agreements with CECED for washing machines in the EU). Some countries even moved from unsuccessful voluntary agreements to MEPS (*e.g.* Brazil). Voluntary agreements can be an effective alternative to minimum energy efficiency standards.

Since they have the support of manufacturers, they can be implemented more rapidly than regulations. Nevertheless, their effectiveness is still dependent on the precise requirements corresponding to genuine additional efforts from industry.

Standards are necessary to remove certain inefficient but inexpensive products from the market, which labelling programmes alone cannot do. They are also needed in areas where the selection criteria of consumers totally exclude energy efficiency (television sets for example), or when the economic stakes for the consumer are very limited. Basically, labelling stimulates technological innovation and the introduction of new more efficient products, while standards effect the gradual removal from the market of the least energy efficient appliances.

Mandatory labelling for several electrical appliances exists in all EU countries based on the same regulations (EU Directives). They include refrigerators and freezers, washing machines, dishwashers and lamps. In OECD Asia and America, about 70% of the countries studied have implemented labels for refrigerators. In Africa, the Middle East and non-OECD Asia, labels are not widespread: they exist for refrigerators in less than 20% of the surveyed countries. Unlike Europe, labels are not always mandatory; however, regulations have proven more effective since they require manufacturers to put labels on all appliances and not just on the most energy efficient ones. Depending on climatic conditions, labelling programmes also concern air conditioners, which are often among the first appliances to be labelled. In most developing countries, second hand appliances account for a large market share of the appliances sold, which reduces the impact of labelling normally restricted to new appliances.

In Europe, about 60% of the countries have standards for refrigerators, which is about the same order of magnitude as in Asia. In OECD America and Asia, a higher proportion of the countries surveyed have such standards (over 80%); in addition, MEPS are imposed on a larger number of appliances (about 12 different types of appliances on average and up to 30 in Canada). Labelling programmes introduced in developing countries are based on the experience of OECD countries and use models that have already been proven: the European label has been used as a model in Brazil, Tunisia, China and Iran, while labels introduced in Thailand and Korea are based on the Australian mode.

Labelling programmes and performance standards are effective instruments, which enable authorities to benefit from low-cost energy savings, consumers to spend less on electricity, and manufacturers to improve their products and become more competitive against imported, less efficient products. As shown by various studies, the increased use of more efficient appliances did not result in a price increase for the consumers, as producers were able to adapt and to benefit from the increased sales.

In Europe, about 60% of the countries have standards for refrigerators, which is about the same order of magnitude as in Asia. In OECD America and Asia, a higher proportion of the countries surveyed have such standards (over 80%); in addition, MEPS are imposed on a larger number of appliances (about 12 different types of appliances on average and up to 30 in Canada).

Labelling programmes introduced in developing countries are based on the experience of OECD countries and use models that have already been proven: the European label has been used as a model in Brazil, Tunisia, China and Iran, while labels introduced in Thailand and Korea are based on the Australian mode.

The European and Australian programmes are considered successful. In the EU for instance, there was a rapid increase in the market share of the most energy efficient appliances. Sales of refrigerators in Class A increased from less than 5% of total sales in 1995 to 23 % in 2000 and 61% in 2005; in addition, 19% of refrigerators sold in 2005 were in the two new more efficient classes (A+ and A++). For washing machines, the progress was even more rapid (1% in 1996, 38% in 2000 and 90% in 2005). Labelling has resulted in market transformation that can be attributed both to the increased interest of consumers in energy efficiency and to changes in the models made available by manufacturers, as well as to other accompanying measures (rebates, information campaigns). The effect of labelling was reinforced by the progressive introduction of MEPS for refrigerators and by the agreement with CECED for washing machines. In anticipation of standards, manufacturers withdrew their less efficient models that had become hard to sell and introduced new more efficient designs to meet new demand and to differentiate themselves from their competitors. The average energy consumption of refrigerators fell from 370 kWh/year in 1990 to around 300 kWh/yr.

In the US, minimum performance standards for the energy efficiency of household appliances also had a large impact. For instance, the average consumption for cold appliances has decreased from 1726 kWh/year in 1972 to 490 kWh today, although this decline has not followed a steady curve. The periods during which energy efficiency ratings improved correspond most to periods when new or reinforced standards were introduced while little or no improvement was observed for the periods in between.

To be effective, labelling programmes and performance standards must be open-ended, *i.e.* regularly revised and upgraded. In the US, changes in the energy efficiency of cold appliances clearly show that energy efficiency improves as a result of new standards but then stabilizes. Faced with new standards, manufacturers adapt the appliances available so that they meet the new minimum requirements, but there are no incentives for them to go beyond what is required if no stricter standards have been planned for the future. For these types of programme, where labels play a secondary role, it is essential to reinforce standards at regular intervals as a way of stimulating technical progress and ensuring a steady improvement in energy efficiency.

In the case of the European and Australian programmes, the balance between energy labels and standards has played a vital role. The requirements are not as strict as they are in the US, but labelling acts as an incentive for manufacturers to differentiate themselves from their competitors and stimulates the introduction of new, more efficient models. However, there is no longer any incentive to innovate when all the models are in the best efficiency classes (Australian experience) or when most of the models on the market have been endorsed with a label (Energy Star programme in the US).

In this respect, the "Top Runner" programme has the particular advantage of making the definition of new targets easier. As the most efficient appliances on the market at a given time are used to set the future standards, there is no need for extensive market or techno-economic analysis to set the minimum energy efficiency standards. With this type of approach, the preparatory work can be shortened and the negotiations between manufacturers and public authorities facilitated as the target corresponds to existing appliances that are already available on the market. Presently, the Top-Runner programme covers 18 energy intensive products, including main household appliances and passenger cars. The savings achieved are impressive: for instance 68% for air conditioners, 55% for refrigerators and 26% for TV.

Generally speaking, manufacturers are opposed to anything that can disrupt market operation, which means efficiency standards in particular, but also labelling systems in certain contexts. Among the arguments frequently advanced by manufacturers is the risk of higher production costs in a context where the possibilities of increasing prices are limited by fierce competition, innovation focused on areas of little importance to consumers, and a less diverse range of products. Experience has shown that such fears are largely unfounded: the turnover and profit levels of manufacturers are not adversely affected by the introduction of standards, nor do the standards compel them to eliminate certain functions to reduce energy consumption. The process of negotiating the introduction of new standards or reinforcing existing ones remains nevertheless conflicting and uncertain.

OTHER REGULATIONS

Other regulations implemented in some countries are mandatory energy consumption reporting, mandatory energy managers, mandatory energy saving plans, mandatory maintenance and obligation of energy savings imposed on utilities. Mandatory audits in industry, as well as energy savings obligations are analysed in detail in another section of the report. Other regulations, not directly linked to energy efficiency, but having significant impact on the energy use (*e.g.* speed limit), are not included in this review.

Energy Consumption Reporting

Some countries have set up regulations requiring designated or large consumers to report their energy consumption, either directly to the government or in their annual report. This measure is seen as an incentive to companies to monitor closely their energy performance. More recently, these measures have also been extended to CO_2 emissions. In India, for example, companies in selected energy intensive sectors in their annual reports to company shareholders provide data on their overall consumption and on the specific energy consumption of manufactured products (*e.g.* cement, pulp, sugar). They also have to provide information on energy saving actions undertaken over the previous year. In some countries, this measure also applies to the building of large public enterprises.

Mandatory Energy Managers

In some countries, there is a regulation requiring the

nomination of an energy manager in companies above a certain size. This concerns about 20% of the countries covered by the survey. This measure usually applies to large consumers in industry (13 countries) and in the service sector (8 countries) (*e.g.* in Denmark for the public sector). In some countries, transport companies are also included (*e.g.* Italy, Portugal, Romania).

Mandatory Energy Saving or DSM Plans

Around 20% of the surveyed countries have set up regulations on the preparation of energy savings plans for consumers, generally in industry (30% in OECD and 10% for non-OECD countries). This measure exists for several sectors, including municipalities in some countries.

Maintenance

Maintenance of energy-consuming equipment is another important field of regulation. The major concern is that without proper maintenance of energy consuming equipment (*e.g.* boilers, vehicles), efficiency will decrease over time: the objective of any regulation is to maintain the initial efficiency of the equipment for as long as possible.

This measure on appliances is mainly in Europe. With the new Directive on buildings, the maintenance of heating boilers will become now mandatory in all EU countries. This measure already existed in Denmark, Italy and Germany. In a few countries (Italy, Romania), regulations on maintenance exist for the transport sector.

The mandatory MOT for cars that exist in many countries may to some extent contribute to saving energy, depending on the aspects to be controlled.

STANDARDS

- *BS EN 572*: Glass in Building–Basic soda lime silicate glass products
  - *Part* 1: 2004 Definitions and general physical and mechanical properties

  - *Part* 2: 2004 Float glass
  - *Part* 3: 2004 Polished wired glass
  - *Part* 4: 2004 Drawn sheet glass
  - *Part* 5: 2004 Patterned glass
  - *Part* 6: 2004 Wired patterned glass
  - *Part* 7: 2004 Wired or unwired channel shaped glass
  - *Part* 8: 2004 Supplied and final cut sizes
  - *Part* 9: 2004 Evaluation of conformity/Product Standard
- *BS 952*: Glass for glazing
  - *Part* 1: 1995 Classification
  - *Part* 2: 1980 Terminology for work on glass
- *BS EN 12600: 2002*: Impact test method and classification for flat glass. This has replaced BS 6206 for glass only.
- *BS 6206: 1981*: Specification for impact performance requirements for flat safety glass and safety plastics for use in buildings-Now only applicable for plastics.
- *BS 6262*: Glazing for buildings
  - *Part* 1: 2005 General methodology for the selection of glazing
  - *Part* 2: 2005 Code of Practice for energy, light and sound
  - *Part* 3: 2005 Code of Practice for fire, security and wind loading
  - *Part* 4: 2005 Code of Practice for safety related to human impact
  - *Part* 5: 2005
  - *Part* 6: 2005 Code of Practice for special applications
  - *Part* 7: 2005 Code of Practice for the provision of information
- *BS 6180: 1999*: Code of practice for barriers in and about buildings
- *BS 5516: 2004*: Patent glazing and sloping glazing for buildings.

  - *Part* 1: 2004 Code for practice for design and installation of sloping and vertical patent glazing
  - *Part* 2: 2005 Code of Practice for sloping glazing
- *BS EN 1063: 2000*: Glass in Building–Security Glazing–Testing and classification of resistance against bullet attack.
- *BS EN 1522: 1999*: Windows, doors, shutters and blinds. Bullet resistance. Requirements and classification.
- *BS 5357: 1995*: Code of practice for installation of security glazing
- *BS EN 356: 2000*: Glass in Building–Security Glazing–Testing and classification of resistance against manual attack. This is the European Standard for anti-bandit glass which will gradually replace BS 5544.
- *BS 5544: 1978*: Specification for anti-bandit glazing (glazing resistant to manual attack). BS EN 356: 2000 is the European Standard which has replaced this standard.
- *BS 6399*: Loading for buildings
  - Part 1: 1996 Code of practice for dead and imposed loads
  - Part 2: 1997 Code of practice for wind loads
- *BS 6340*: Shower units
  - Part 1: 1985 Specification for prefabricated shower enclosures and shower cabinets
- BS EN 14072:2003: Glass in furniture test methods BS 7376:2004: Specification for inclusion of glass in the construction of tables or trolleys
- BS 7449:1991: Specification for inclusion of glass in the construction of furniture, other than tables or trolleys including cabinets, shelving systems and wall hung or free standing mirrors.
- BS EN 12354-3
- BS EN ISO 717: Rating of sound insulation in buildings and of building elements
  - *Part* 1: 1997: Airborne sound insulation
  - *Part* 2: 1997: Impact sound insulation

- BS 5821: Methods of rating the sound insulation in building and of building elements
  - *Part* 3: 1984 Methods for rating the airborne sound insulation of façade elements and façades
- BS 8213: Safe cleaning of windows
- BS EN 1279: Glass in buildings. Insulating Glass Units.
  - *Part* 1: 2004 Generalities, dimensional tolerances and rules for the system description.
  - *Part* 2: 2005 Long term test method and requirements for moisture penetration
  - *Part* 3: 2005 Long term test method and requirements for gas leakage rate and for gas concentration tolerances.
  - *Part* 4: 2002 Methods of test for the physical attributes of edge seals
  - *Part* 5: 2005 Evaluation of conformity.
  - *Part* 6: 2002 Factory production control and periodic tests.
- BS EN 673: 1998: Thermal performance of windows and doors. Determination of thermal transmittance by hot box method.
- BS 6993: Thermal and radiometric properties of glazing
  - *Part* 2: 1990 Method for direct measurement of U value (Thermal Transmittance)
- BS 874: Methods for determining thermal insulating properties
  - *Part* 1: Section 3.1: 1987 Guarded hot-box method
  - *Part* 2: Section 3.2: 1990 Calibrated hot-box method
- BS EN 12567: Glass in Building–Determination of thermal transmittance (U value) –Measurement and testing standard
  - *Part* 1: 2000 Complete windows and other projecting windows.
  - *Part* 2: 2005 Roof windows and other projecting windows.
- BS EN 410: 1998: Glass in Building–Determination of luminous and solar characteristics of glazing

- BS 476: Fire tests on building materials and structures
  - *Part* 3: 2004 Classification and method of test for external fire exposure to roofs.
  - *Part* 4: 1970 (1984) Non-combustibility test for materials
  - *Part* 6: 1989 Method of test for fire propagation for products
  - *Part* 7: 1997 Method for classification of the surface spread of flame of products
  - *Part* 10: 1983 Guide to the principles and application of fire testing.
  - *Part* 11: 1982 Method for assessing the heat emission from building materials.
  - *Part* 12: 1991 Method of test for ignitability of products by direct flame impingement.
  - *Part* 13: 1987, ISO 5657: 1986 Method of measuring the ignitability of products subjected to thermal irradiance.
  - *Part* 15: 1993, ISO -1:1993 Method for measuring the rate of heat release products.
  - *Part* 20: 1987 Method for the determination of the fire resistance of elements of construction (general principles)
  - *Part* 21: 1987 Method for the determination of the fire resistance of loadbearing elements of construction
  - *Part* 22: 1987 Methods for the determination of the fire resistance of non-loadbearing elements of construction.
  - *Part* 23: 1987 Methods for the determination of the contribution of components to the fire resistance of a structure.
  - *Part* 24: 1987, ISO 6944: 1985 Method for the determination of the fire resistance of ventilation ducts.
  - *Part* 31.1: 1983 Methods for measuring smoke penetration through doorsets and shutter

assemblies. Method of measurement under ambient temperature conditions.
  - *Part* 32: 1989 Guide to full scale fire tests within buildings.
  - *Part* 33: 1993, ISO 9705: 1993 Full-scale room test for surface products

These standards will be replaced by a number of European Standards over the next 5-10 years.

These will be:

- BS EN 1363: 1999 Fire Resistance tests
  - *Part* 1 General requirements.
  - *Part* 2 Alternative and additional procedures.
- BS EN 1364: Part 1–walls including glazing.
- BS EN 1634: 2000 Fire resistance tests for door and shutter assemblies.
  - *Part* 1: 2000 Fire doors and shutters.
  - *Part* 3: 2004 Smoke control doors and shutters.

This will bring in 3 classifications for fire resistant products:

- E – Integrity
- EW – Reduced Heat Radiation
- EI – Insulation
- BS EN 14600: 2005 Doorsets and openable windows with fire resisting and/or smoke control characteristics. Requirements and classification.
- BS 5588: Fire precautions in the design, construction and use of buildings
  - *Part* 0: 1996 Guide to fire safety codes of practice for particular premises/applications
  - *Part* 1: 1990 Code of Practice for residential buildings
  - *Part* 2: 1985 Code of Practice for shops
  - *Part* 3: 1983 Code of Practice for office buildings
  - *Part* 4: 1978 Code of Practice for smoke control in protected escape routes using pressurization.
  - *Part* 5: 2004 Access and facilities for fire-fighting.
  - *Part* 6: 1991 Code of Practice for places of assembly.

  - *Part* 7: 1997 Code of Practice for the incorporation of atria in buildings.
  - *Part* 8: 1999 Code of Practice for means of escape for disabled people.
  - *Part* 9: 1999 Code of Practice for ventilation and air conditioning ductwork
  - *Part* 10: 1991 Code of Practice for shopping complexes.
  - *Part* 11: 1997 Code of Practice for shops,offices, industrial, storage and other similar buildings.
  - *Part* 12: 2004 Managing fire safety
- BS EN 357: 2004 Glass in building. Fire resistant glazed elements with transparent or translucent glass products. Classification of fire resistance.
- BS 8000: Workmanship on building sites
  - *Part* 7: 1990 Code of practice for glazing
- EN 12150-1 Thermally toughened soda lime silicate safety glass
  - *Part* 1: Definition and description
  - *Part* 2: Evaluation of conformity/Product standard
- EN ISO 12543-1 Laminated glass and laminated safety glass
  - *Part* 1: Definitions and description of component parts
  - *Part* 2: Laminated safety glass
  - *Part* 3: Laminated glass
  - *Part* 4: Test methods for durability
  - *Part* 5: Dimensions and edge finishing
  - *Part* 6: Appearance
- EN 14449 Laminated glass and laminated safety glass– Evaluation of conformity/Product standard
- EN 1096-1 Coated glass
  - *Part* 1: Definitions and classification
  - *Part* 2: Requirements and tests method for class A, B and S coatings
  - *Part* 3: Requirements and test methods for class C and D coatings

  - *Part* 4: Evaluation of conformity/Product Standard
- EN 1036-1 Mirrors from silver coated float glass for internal use
- EN 1036-2: 2009 Mirrors from silver coated float glass for internal use
  - *Part* 2: Evaluation of conformity/Product Standard
- EN 1863-1 Heat strengthened soda lime silicate glass
  - *Part* 1: Definition and description
  - *Part* 2: Evaluation of conformity/Product standard
- EN 14179-1 Heat soaked thermally toughened soda lime silicate safety glass
  - *Part* 1: Definition and description
  - *Part* 2: Evaluation of conformity/Product standard
- EN 12898: Determination of the emissivity
- EN ISO 14438: Determination of energy balance-Calculation method
- EN 12758: Glazing and airborne sound insulation-Product descriptions and determination of properties
- EN 13501-2: Fire classification of construction products and building elements–Part 2: Classification using data from fire resistance tests, excluding ventilation services.
- EN 13541: Security glazing-Testing and classification of resistance against explosion

4

Designing and Planning

DESIGN AND CONSTRUCTION AS AN INTEGRATED SYSTEM

In the planning of facilities, it is important to recognize the close relationship between design and construction. These processes can best be viewed as an integrated system. Broadly speaking, design is a process of creating the description of a new facility, usually represented by detailed plans and specifications; construction planning is a process of identifying activities and resources required to make the design a physical reality. Hence, construction is the implementation of a design envisioned by architects and engineers. In both design and construction, numerous operational tasks must be performed with a variety of precedence and other relationships among the different tasks.

Several characteristics are unique to the planning of constructed facilities and should be kept in mind even at the very early stage of the project life cycle.

These include the following:

- Nearly every facility is custom designed and constructed, and often requires a long time to complete.

- Both the design and construction of a facility must satisfy the conditions peculiar to a specific site.
- Because each project is site specific, its execution is influenced by natural, social and other locational conditions such as weather, labor supply, local building codes, etc.
- Since the service life of a facility is long, the anticipation of future requirements is inherently difficult.
- Because of technological complexity and market demands, changes of design plans during construction are not uncommon.

In an integrated system, the planning for both design and construction can proceed almost simultaneously, examining various alternatives which are desirable from both viewpoints and thus eliminating the necessity of extensive revisions under the guise of value engineering. Furthermore, the review of designs with regard to their constructibility can be carried out as the project progresses from planning to design. However, if the design professionals are expected to assume such responsibilities, they must be rewarded for sharing the risks as well as for undertaking these additional tasks. Similarly, when construction contractors are expected to take over the responsibilities of engineers, such as devising a very elaborate scheme to erect an unconventional structure, they too must be rewarded accordingly. As long as the owner does not assume the responsibility for resolving this risk-reward dilemma, the concept of a truly integrated system for design and construction cannot be realised.

It is interesting to note that European owners are generally more open to new technologies and to share risks with designers and contractors. In particular, they are more willing to accept responsibilities for the unforeseen subsurface conditions in geotechnical engineering. Consequently, the designers and contractors are also more willing to introduce new techniques in order to reduce the time and cost of construction. In European practice, owners typically present contractors with a conceptual design, and contractors prepare detailed designs, which are checked by the owner's engineers. Those detailed designs may be alternate designs, and specialty contractors may also prepare detailed alternate designs.

RESPONSIBILITY FOR SHOP DRAWINGS

The willingness to assume responsibilities does not come easily from any party in the current litigious climate of the construction industry in the United States. On the other hand, if owner, architect, engineer, contractor and other groups that represent parts of the industry do not jointly fix the responsibilities of various tasks to appropriate parties, the standards of practice will eventually be set by court decisions. In an attempt to provide a guide to the entire spectrum of participants in a construction project, the American Society of Civil Engineers issued a Manual of Professional Practice entitled *Quality in the Constructed Project* in 1990. This manual is intended to help bring a turn around of the fragmentation of activities in the design and construction process.

However, since the responsibility for preparing shop drawings was traditionally assigned to construction contractors, design professionals took the view that the review process was advisory and assumed no responsibility for their accuracy. This justification was ruled unacceptable by a court in connection with the walkway failure at the Hyatt Hotel in Kansas City in 1985. In preparing the ASCE Manual of Professional Practice for Quality in the Constructed Project, the responsibilities for preparation of shop drawings proved to be the most difficult to develop. The reason for this situation is not difficult to fathom since the responsibilities for the task are diffused, and all parties must agree to the new responsibilities assigned to each in the recommended risk-reward relations shown in Table.

Traditionally, the owner is not involved in the preparation and review of shop drawings, and perhaps is even unaware of any potential problems. In the recommended practice, the owner is required to take responsibility for providing adequate time and funding, including approval of scheduling, in order to allow the design professionals and construction contractors to perform satisfactorily.

Table Recommended Responsibility for Shop Drawings

Task	Responsible Party		
	Owner	Design Professional	Construction Contractor
Provide adequate time and funding for shop drawing preparation and review	Prime		
Arrange for structural design	Prime		
Provide structural design		Prime	
Establish overall responsibility for connection design		Prime	
Accomplish connection design (by design professional)		Prime	
Alternatively, provide loading requirement and other information necessary for shop drawing preparation		Prime	
Alternatively, accomplish some or all of connection design (by constuctor with a licensed P.E.)			Prime
Specify shop drawing requirements and procedures	Review	Prime	
Approve proper scheduling	Prime	Assisting	Assisting
Provide shop drawing and submit the drawing on schedule			Prime
Make timely reviews and approvals		Prime	
Provide erection procedures, construction bracing, shoring, means, methods and techniques of construction, and construction safety			Prime

MODEL METRO PROJECT

Under Italian law, unforeseen subsurface conditions are the owner's responsibility, not the contractor's. This is a striking difference from U.S. construction practice where changed conditions clauses and claims and the adequacy of prebid site investigations are points of contention. In effect, the Italian law means that the owner assumes those risks. But under the same law, a contractor may elect to assume the risks in order to lower the bid price and thereby beat the competition.

According to the Technical Director of Rodio, the Milan-based contractor which is heavily involved in the grouting job for tunneling in the Model Metro project in Milan, Italy, there are two typical contractual arrangements for specialized subcontractor firms such as theirs. One is to work on a unit price basis with no responsibility for the design. The other is what he calls the "nominated subcontractor" or turnkey method: prequalified subcontractors offer their own designs and guarantee the price, quality, quantities, and, if they wish, the risks of unforeseen conditions. At the beginning of the Milan metro project, the Rodio contract ratio was 50/50 unit price and turnkey. The firm convinced the metro owners that they would save money with the turnkey approach, and the ratio became 80% turnkey. What's more, in the work packages where Rodio worked with other grouting specialists, those subcontractors paid Rodio a fee to assume all risks for unforeseen conditions.

Under these circumstances, it was critical that the firm should know the subsurface conditions as precisely as possible, which was a major reason why the firm developed a computerized electronic sensing programme to predict stratigraphy and thus control grout mixes, pressures and, most important, quantities.

INNOVATION AND TECHNOLOGICAL FEASIBILITY

The planning for a construction project begins with the generation of concepts for a facility which will meet market demands and owner needs. Innovative concepts in design are highly valued not for their own sake but for their contributions to reducing costs and to the improvement of aesthetics, comfort or convenience as embodied in a well-designed facility. However, the constructor as well as the design professionals must have an appreciation and full understanding of the technological complexities often associated with innovative designs in order to provide a safe and sound facility. Since these concepts are often preliminary or tentative, screening studies are carried out to determine the overall technological viability and economic attractiveness without pursuing these concepts in great detail. Because of the ambiguity of the objectives and the uncertainty of external events, screening studies call for uninhibited innovation in creating new concepts and judicious judgment in selecting the appropriate ones for further consideration. One of the most important aspects of design innovation is the necessity of communication in the design/ construction partnership. In the case of bridge design, it can be illustrated by the following quotation from Lin and Gerwick concerning bridge construction:

The great pioneering steel bridges of the United States were built by an open or covert alliance between designers and constructors. The turnkey approach of designer-constructor has developed and built our chemical plants, refineries, steel plants, and nuclear power plants. It is time to ask, seriously, whether we may not have adopted a restrictive approach by divorcing engineering and construction in the field of bridge construction.

If a contractor-engineer, by some stroke of genius, were to present to design engineers today a wonderful new scheme for long span prestressed concrete bridges that made them far cheaper, he would have to make these ideas available to all other constructors, even limiting or watering them down so as to "get a group of truly competitive bidders." The engineer would have to make sure that he found other contractors to bid against the ingenious innovator. If an engineer should, by a similar stroke of genius, hit on such a unique and brilliant scheme, he would have to worry, wondering if the low bidder would be one who had any concept of what he was trying to accomplish or was in any way qualified for high class technical work. Innovative design concepts must be tested for technological feasibility. Three levels of technology are of special concern: technological requirements for operation or production, design resources and construction technology. The first refers to the new technologies that may be introduced in a facility which is used for a certain type of production such as chemical processing or nuclear power generation. The second refers to the design capabilities that are available to the designers, such as new computational methods or new materials. The third refers to new technologies which can be adopted to construct the facility, such as new equipment or new construction methods. A new facility may involve complex new technology for operation in hostile environments such as severe climate or restricted accessibility. Large projects with unprecedented demands for resources such as labor supply, material and infrastructure may also call for careful technological feasibility studies.

Major elements in a feasibility study on production technology should include, but are not limited to, the following:

- Project type as characterized by the technology required, such as synthetic fuels, petrochemicals, nuclear power plants, etc.
- Project size in dollars, design engineer's hours, construction labor hours, etc.
- Design, including sources of any special technology which require licensing agreements.
- Project location which may pose problems in environmental protection, labor productivity and special risks.

An example of innovative design for operation and production is the use of entropy concepts for the design of integrated chemical processes. Simple calculations can be used to indicate the minimum energy requirements and the least number of heat exchange units to achieve desired objectives. The result is a new incentive and criterion for designers to achieve more effective designs. Numerous applications of the new methodology has shown its efficacy in reducing both energy costs and construction expenditures. This is a case in which innovative design is not a matter of trading-off operating and capital costs, but better designs can simultaneously achieve improvements in both objectives. The choice of construction technology and method involves both *strategic* and *tactical* decisions about appropriate technologies and the best sequencing of operations. For example, the extent to which prefabricated facility components will be used represents a *strategic* construction decision. In turn, prefabrication of components might be accomplished off-site in existing manufacturing facilities or a temporary, on-site fabrication plant might be used. Another example of a strategic decision is whether to install mechanical equipment in place early in the construction process or at an intermediate stage. Strategic decisions of this sort should be integrated with the process of facility design in many cases. At the tactical level, detailed decisions about how to accomplish particular tasks are required, and such decisions can often be made in the field.

Construction planning should be a major concern in the development of facility designs, in the preparation of cost estimates, and in forming bids by contractors. Unfortunately, planning for the construction of a facility is often treated as an after thought by design professionals. This contrasts with manufacturing practices in which the *assembly* of devices is a major concern in design. Design to insure ease of assembly or construction should be a major concern of engineers and architects. As the Business Roundtable noted, "All too often chances to cut schedule time and costs are lost because construction operates as a production process separated by a chasm from financial planning, scheduling, and engineering or architectural design. Too many engineers, separated from field experience, are not up to date about how to build what they design, or how to design so structures and equipment can be erected most efficiently."

INNOVATIVE USE OF STRUCTURAL FRAMES FOR BUILDINGS

The structural design of skyscrapers offers an example of innovation in overcoming the barrier of high costs for tall buildings by making use of new design capabilities. A revolutionary concept in skyscraper design was introduced in the 1960's by Fazlur Khan who argued that, for a building of a given height, there is an appropriate structural system which would produce the most efficient use of the material. Before 1965, most skyscrapers were steel rigid frames. However, Fazlur

Khan believed that it was uneconomical to construct all office buildings of rigid frames, and proposed an array of appropriate structural systems for steel buildings of specified heights as shown in Figure 4.1. By choosing an appropriate structural system, an engineer can use structural materials more efficiently. For example, the 60-story Chase Manhattan Building in New York used about 60 pounds per square foot of steel in its rigid frame structure, while the 100-story John Hancock Center in Chicago used only 30 pounds per square foot for a trusted tube system. At the time the Chase Manhattan Building was constructed, no bracing was used to stiffen the core of a rigid frame building because design engineers did not have the computing tools to do the complex mathematical analysis associated with core bracing.

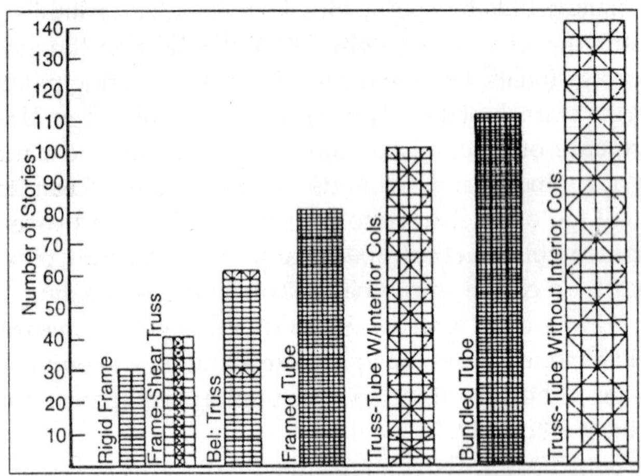

Fig. 4.1 Proposed Structural System Fir Steel Buildings

INNOVATION AND ECONOMIC FEASIBILITY

Innovation is often regarded as the engine which can introduce construction economies and advance labor productivity. This is obviously true for certain types of innovations in industrial production technologies, design capabilities, and construction equipment and methods. However, there are also limitations due to the economic infeasibility of such innovations, particularly in the segments of construction industry which are more fragmented and permit ease of entry, as in the construction of residential housing.

Market demand and firm size play an important role in this regard. If a builder is to construct a larger number of similar units of buildings, the cost per unit may be reduced. This relationship between the market demand and the total cost of production may be illustrated schematically as in Figure 4.2. An initial threshold or fixed cost F is incurred to allow any production. Beyond this threshold cost, total cost increases faster than the units of output but at a decreasing rate. At each point on this total cost curve, the average cost is represented by the slope of a line from the origin to the point on the curve. At a point H, the average cost per unit is at a minimum. Beyond H to the right, the total cost again increases faster than the units of output and at an increasing rate. When the rate of change of the average cost slope is decreasing or constant as between 0 and H on the curve, the range between 0 and H is said to be *increasing return to scale*; when the rate of change of the average cost slope is increasing as beyond H to the right, the region is said to be *decreasing return to scale*. Thus, if fewer than h units are constructed, the unit price will be higher than that of exactly h units. On the other hand, the unit price will increase again if more than h units are constructed.

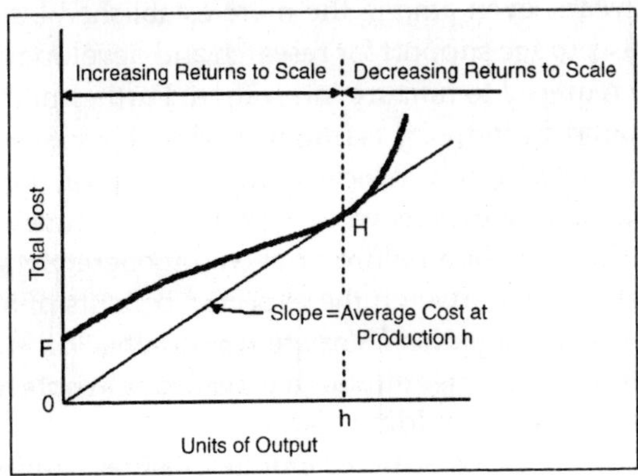

Fig. 4.2 Market Demand and Total Cost Relationship

Nowhere is the effect of market demand and total cost more evident than in residential housing. The housing segment in the last few decades accepted many innovative technical improvements in building materials which were promoted by material suppliers. Since material suppliers provide products to a large number of homebuilders and others, they are in a better position to exploit production economies of scale and to support new product development. However, homebuilders themselves have not been as successful in making the most fundamental form of innovation which encompasses changes in the technological process of homebuilding by shifting the mixture of labor and material inputs, such as substituting large scale off-site prefabrication for on-site assembly.

There are several major barriers to innovation in the technological process of homebuilding, including demand instability, industrial fragmentation, and building codes. Since market demand for new homes follows demographic trends and other socio-economic conditions, the variation in home building has been anything but regular. The profitability of the homebuilding industry has closely matched aggregate output levels. Since entry and exist from the industry are relatively easy, it is not uncommon during periods of slack demand to find builders leaving the market or suspending their operations until better times. The inconsistent levels of retained earnings over a period of years, even among the more established builders, are likely to discourage support for research and development efforts which are required to nurture innovation. Furthermore, because the homebuilding industry is fragmented with a vast majority of homebuilders active only in local regions, the typical homebuilder finds it excessively expensive to experiment with new designs. The potential costs of a failure or even a moderately successful innovation would outweigh the expected benefits of all but the most successful innovations. Variation in local building codes has also caused inefficiencies although repeated attempts have been made to standardize building codes.

In addition to the scale economies visible within a sector of the construction market, there are also possibilities for scale economies in individual facility. For example, the relationship between the size of a building (expressed in square feet) and the input labor (expressed in laborhours per square foot) varies for different types and sizes of buildings.

As shown in Figures 4.3, these relationships for several types of buildings exhibit different characteristics. The labor hours per square foot decline as the size of facility increases for houses, public housing and public buildings. However, the labor hours per square foot almost remains constant for all sizes of school buildings and increases as the size of a hospital facility increases.

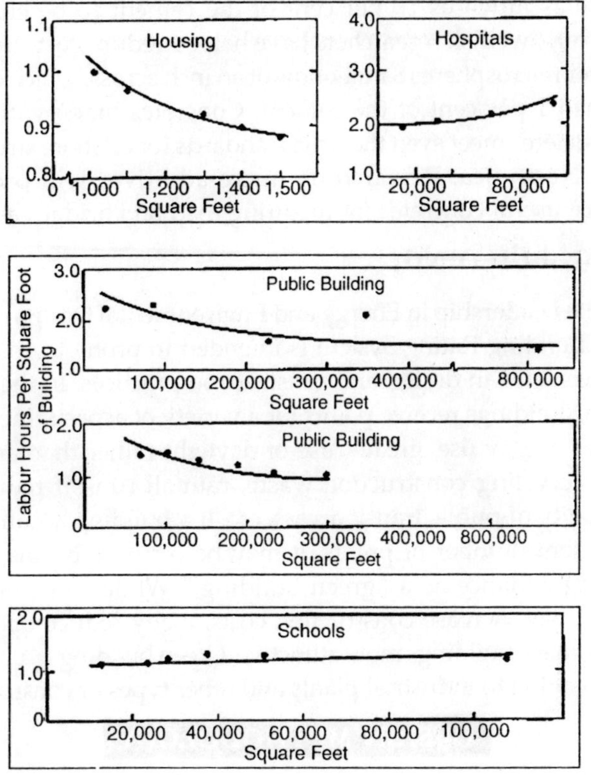

Fig. 4.3 Illustrative Relationships between Building Size and Input Labor by Types of Building

USE OF NEW MATERIALS

In recent years, an almost entirely new set of materials is emerging for construction, largely from the aerospace and electronics industries. These materials were developed from new knowledge about the structure and properties of materials as well as new techniques for altering existing materials. Additives to traditional materials such as concrete and steel are particularly prominent. For example, it has been known for some time that polymers would increase concrete strength, water resistance and ability to insulate when they are added to the cement. However, their use has been limited by their costs since they have had to replace as much as 10 per cent of the cement to be effective. However, Swedish researchers have helped reduce costs by using polymer microspheres 8 millionths of an inch across, which occupy less than 1 per cent of the cement. Concretes made with these microspheres meet even the strict standards for offshore structures in the North Sea. Research on micro-additives will probably produce useful concretes for repairing road and bridges as well.

GREEN BUILDINGS

The Leadership in Energy and Environmental Design (LEED) Green Building Rating System is intended to promote voluntary improvements in design and construction practices. In the rating system, buildings receive points for a variety of aspects, including reduced energy use, greater use of daylight rather than artificial lights, recycling construction waste, rainfall run-off reduction, availability of public transit access, etc. If a building accumulates a sufficient number of points, it may be certified by the Green Building Alliance as a "green building." While some of these aspects may increase construction costs, many reduce operating costs or make buildings more attractive. Green building approaches are spreading to industrial plants and other types of construction.

DESIGN METHODOLOGY

While the conceptual design process may be formal or informal, it can be characterized by a series of actions: formulation, analysis, search, decision, specification, and modification. However, at the early stage in the development of a new project, these actions are highly interactive as illustrated in Figure 4.4. Many iterations of redesign are expected to refine the functional requirements, design concepts and financial constraints, even though the analytic tools applied to the solution of the problem at this stage may be very crude.

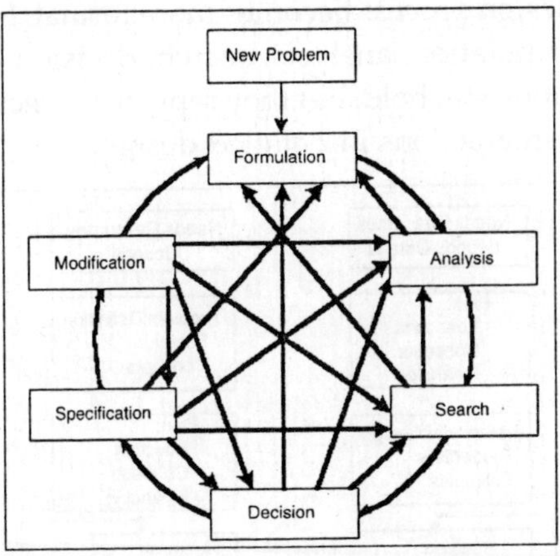

Fig. 4.4 Conceptual Design Process

The series of actions taken in the conceptual design process may be described as follows:

- Formulation refers to the definition or description of a design problem in broad terms through the synthesis of ideas describing alternative facilities.
- Analysis refines the problem definition or description by separating important from peripheral information and by pulling together the essential detail. Interpretation and prediction are usually required as part of the analysis.
- Search involves gathering a set of potential solutions for performing the specified functions and satisfying the user requirements.
- Decision means that each of the potential solutions is evaluated and compared to the alternatives until the best solution is obtained.

- Specification is to describe the chosen solution in a form which contains enough detail for implementation.
- Modification refers to the change in the solution or re-design if the solution is found to be wanting or if new information is discovered in the process of design.

As the project moves from conceptual planning to detailed design, the design process becomes more formal. In general, the actions of formulation, analysis, search, decision, specification and modification still hold, but they represent specific steps with less random interactions in detailed design.

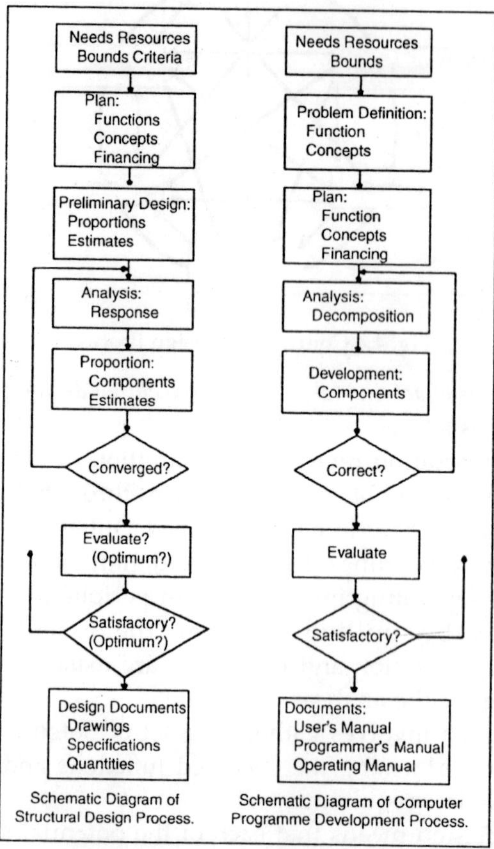

Fig. 4.5 An Analogy between Structural Design and Computer Programme Development Process

The design methodology thus formalized can be applied to a variety of design problems. For example, the analogy of the schematic diagrams of the structural design process and of the computer programme development process is shown in Figure 4.5.

The basic approach to design relies on decomposition and integration. Since design problems are large and complex, they have to be decomposed to yield subproblems that are small enough to solve. There are numerous alternative ways to decompose design problems, such as decomposition by functions of the facility, by spatial locations of its parts, or by links of various functions or parts. Solutions to subproblems must be integrated into an overall solution. The integration often creates conceptual conflicts which must be identified and corrected.

A hierarchical structure with an appropriate number of levels may be used for the decomposition of a design problem to sub problems. For example, in the structural design of a multistory building, the building may be decomposed into floors, and each floor may in turn be decomposed into separate areas. Thus, a hierarchy representing the levels of building, floor and area is formed.

Different design styles may be used. The adoption of a particular style often depends on factors such as time pressure or available design tools, as well as the nature of the design problem.

Examples of different styles are:

- *Top-down design*: Begin with a behaviour description of the facility and work towards descriptions of its components and their interconnections.
- *Bottom-up design*: Begin with a set of components, and see if they can be arranged to meet the behaviour description of the facility.

The design of a new facility often begins with the search of the files for a design that comes as close as possible to the one needed. The design process is guided by accumulated experience and intuition in the form of heuristic rules to find acceptable solutions. As more experience is gained for this particular type of facility, it often becomes evident that parts of the design problem are amenable to rigorous definition and algorithmic solution. Even formal optimization methods may be applied to some parts of the problem.

FUNCTIONAL DESIGN

The objective of functional design for a proposed facility is to treat the facility as a complex system of interrelated spaces which are organized systematically according to the functions to be performed in these spaces in order to serve a collection of needs. The arrangement of physical spaces can be viewed as an iterative design process to find a suitable floor plan to facilitate the movement of people and goods associated with the operations intended.

A designer often relies on a heuristic approach, *i.e.*, applying selected rules or strategies serving to stimulate the investigation in search for a solution.

The heuristic approach used in arranging spatial layouts for facilities is based generally on the following considerations:

- Identification of the goals and constraints for specified tasks,
- Determination of the current state of each task in the iterative design process,
- Evaluation of the differences between the current state and the goals,
- Means of directing the efforts of search towards the goals on the basis of past experience.

Hence, the procedure for seeking the goals can be recycled iteratively in order to make tradeoffs and thus improve the solution of spatial layouts.

Consider, for example, an integrated functional design for a proposed hospital. Since the responsibilities for satisfying various needs in a hospital are divided among different groups of personnel within the hospital administrative structure, a hierarchy of functions corresponding to different levels of responsibilities is proposed in the systematic organization of hospital functions. In this model, the functions of a hospital system are decomposed into a hierarchy of several levels:

- *Hospital*: Conglomerate of all hospital services resulting from top policy decisions,
- *Division*: Broadly related activities assigned to the same general area by administrative decisions,
- *Department*: Combination of services delivered by a service or treatment group,
- *Suite*: Specific style of common services or treatments performed in the same suite of rooms,
- *Room*: All activities that can be carried out in the same internal environment surrounded by physical barriers,
- *Zone*: Several closely related activities that are undertaken by individuals,
- *Object*: A single activity associated with an individual.

In the integrated functional design of hospitals, the connection between physical spaces and functions is most easily made at the lowest level of the hierarchy, and then extended upward to the next higher level. For example, a bed is a physical object immediately related to the activity of a patient. A set of furniture consisting of a bed, a night table and an armchair arranged comfortably in a zone indicates the sphere of private activities for a patient in a room with multiple occupancy.

Thus, the spatial representation of a hospital can be organized in stages starting from the lowest level and moving to the top. In each step of the organization process, an element (space or function) under consideration can be related directly to the elements at the levels above it, to those at the levels below it, and to those within the same level.

Since the primary factor relating spaces is the movement of people and supplies, the objective of arranging spaces is the minimization of movement within the hospital. On the other hand, the internal environmental factors such as atmospheric conditions (pressure, temperature, relative humidity, odour and particle pollution), sound, light and fire protection produce constraining effects on the arrangement of spaces since certain spaces cannot be placed adjacent to other spaces because of different requirements in environmental conditions. The consideration of logistics is important at all levels of the hospital system. For example, the travel patterns between objects in a zone or those between zones in a room are frequently equally important for devising an effective design. On the other hand, the adjacency desirability matrix based upon environmental conditions will not be important for organization of functional elements below the room level since a room is the lowest level that can provide a physical barrier to contain desirable environmental conditions. Hence, the organization of functions for a new hospital can be carried out through an interactive process, starting from the functional elements at the lowest level that is regarded as stable by the designer, and moving step by step up to the top level of the hierarchy.

Due to the strong correlation between functions and the physical spaces in which they are performed, the arrangement of physical spaces for accommodating the functions will also follow the same iterative process. Once a satisfactory spatial arrangement is achieved, the hospital design is completed by the selection of suitable building components which complement the spatial arrangement.

TOP-DOWN DESIGN STYLE

In the functional design of a hospital, the designer may begin with a "reference model", *i.e.* the spatial layouts of existing hospitals of similar size and service requirements. On the basis of past experience, spaces are allocated to various divisions as shown schematically in Figure 4.6. The space in each division is then divided further for various departments in the division, and all the way down the line of the hierarchy. In every step along the way, the pertinent information of the elements immediately below the level under consideration will be assessed in order to provide input for making necessary adjustments at the current level if necessary. The major drawback of the top-down design style is that the connection between physical spaces and functions at lower levels cannot be easily anticipated. Consequently, the new design is essentially based on the intuition and experience of the designer rather than an objective analysis of the functions and space needs of the facility. Its greatest attraction is its simplicity which keeps the time and cost of design relatively low.

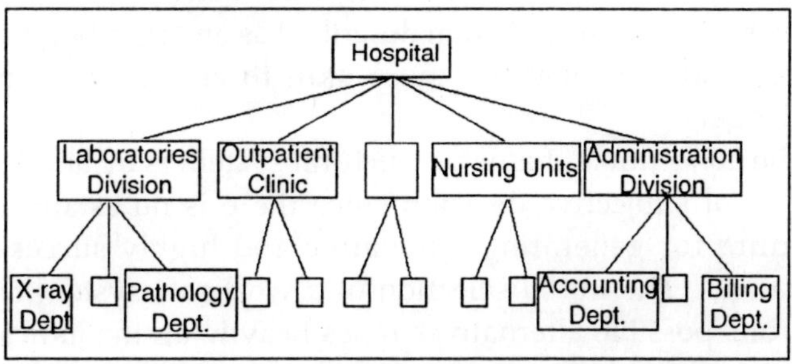

Fig. 4.6 A Model for Top-Down Design of a Hospital

BOTTOM-UP DESIGN STYLE

A multi-purpose examination suite in a hospital is used as an illustration of bottom-up design style. In Figure 4.7, the most basic elements (furniture) are first organized into zones which make up the room. Thus the size of the room is determined by spatial layout required to perform the desired services. Finally, the suite is defined by the rooms which are parts of the multi-purpose examination suite.

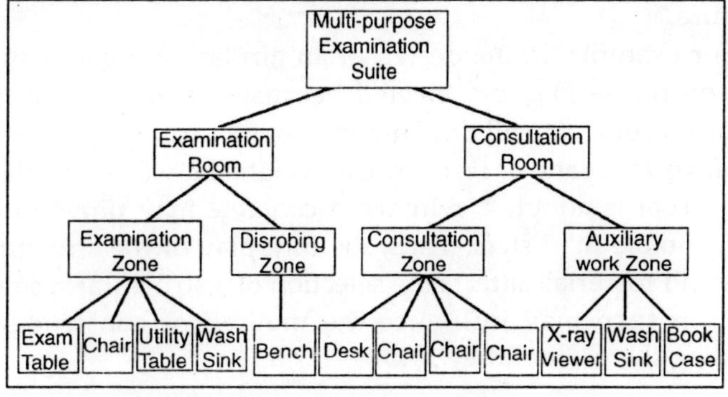

Fig. 4.7 A Model for Bottom-up Design of an Examination Suite

PHYSICAL STRUCTURES

The structural design of complex engineering systems generally involves both synthesis and analysis. Synthesis is an inductive process while analysis is a deductive process. The activities in synthesis are often described as an art rather than a science, and are regarded more akin to creativity than to knowledge.

The conception of a new structural system is by and large a matter of subjective decision since there is no established procedure for generating innovative and highly successful alternatives. The initial selection of a workable system from numerous possible alternatives relies heavily on the judicious judgment of the designer.

Once a structural system is selected, it must be subjected to vigorous analysis to insure that it can sustain the demands in its environment. In addition, compatibility of the structural system with mechanical equipment and piping must be assured.

For traditional types of structures such as office buildings, there are standard systems derived from the past experience of many designers. However, in many situations, special systems must be developed to meet the specified requirements. The choice of materials for a structure depends not only on the suitability of materials and their influence on the form of the structure.

For example, in the design of an airplane hangar, a steel skeleton frame may be selected because a similar frame in reinforced concrete will limit the span of the structure owing to its unfavorable ratio or resistance to weight. However, if a thin-shelled roof is adopted, reinforced concrete may prove to be more suitable than steel. Thus, the interplay of the structural forms and materials affects the selection of a structural system, which in turn may influence the method of construction including the use of falsework.

STEEL FRAME SUPPORTING A TURBO-BLOWER

The design of a structural frame supporting a turbo-blower

supplying pressurized air to a blast furnace in a steel mill can be used to illustrate the structural design process. As shown in Figure 4.8, the turbo-blower consists of a turbine and a blower linked to an air inlet stack.

Since the vibration of the turbo-blower is a major concern to its operation, a preliminary investigation calls for a supporting frame which is separated from the structural frame of the building.

An analysis of the vibration characteristics of the turbo-blower indicates that the lowest mode of vibration consists of independent vibration of the turbine shaft and the blower shaft, with higher modes for the coupled turbo-blower system when both shafts vibrate either in-phase or out-of-phase.

Consequently, a steel frame with separate units for the blower side and the turbine side is selected. The columns of the steel frame are mounted on pile foundation and all joints of the steel frame are welded to reduce the vibration levels.

Since the structural steel frame also supports a condenser, an air inlet and exhaust, and a steam inlet and exhaust in addition to the turbo-blower, a static analysis is made to size its members to support all applied loads. Then, a dynamic analysis is conducted to determine the vibration characteristics of the system incorporating the structural steel frame and the turbo-blower. When the limiting conditions for static loads and natural frequencies of vibration are met, the design is accepted as satisfactory.

Turbo Blower In the chemistry of plating, and metal finishing, agitation renews the cathode film, decreases polarization and enables the use of higher current density. This means finer grain deposits and higher plating speeds. In rinsing, agitation prevents localized concentration of drag-in.

Multi Suction Turbo Blower Low Pressure and High Volumetric Suction Blower These are also high volumetric but low pressure blowers having 50 mm to 200 mm wg pressure and 2000 cfm to 16000 cfm volume capacity which are used to suck the Fumes, sometimes are called fume exhauster.

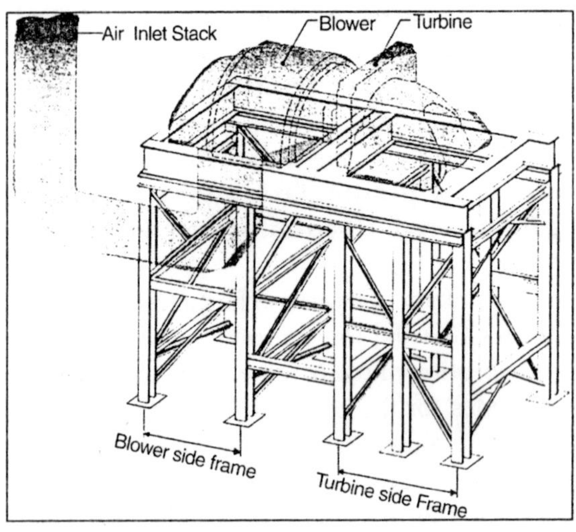

Fig. 4.8 Steel Frame Supporting a Turbo-Blower

MULTIPLE HIERARCHY DESCRIPTIONS OF PROJECTS

In the previous section, a hierarchy of functional spaces was suggested for describing a facility. This description is appropriate for functional design of spaces and processes within a building, but may be inadequate as a view of the facility's structural systems.

A hierarchy suitable for this purpose might divide elements into *structural functions* such as slabs, walls, frames, footings, piles or mats. Lower levels of the hierarchy would describe individual design elements.

For example, frames would be made up of column, beam and diagonal groups which, in turn, are composed of individual structural elements. These individual structural elements comprise the limits on functional spaces such as rooms in a different hierarchical perspective. Designers typically will initiate a view appropriate for their own concerns, and these different hierarchical views must be synthesized to insure consistency and adequacy of the overall design.

GEOTECHNICAL ENGINEERING INVESTIGATION

Since construction is site specific, it is very important to investigate the subsurface conditions which often influence the design of a facility as well as its foundation. The uncertainty in the design is particularly acute in geotechnical engineering so that the assignment of risks in this area should be a major concern. Since the degree of uncertainty in a project is perceived differently by different parties involved in a project, the assignment of unquantifiable risks arising from numerous unknowns to the owner, engineer and contractor is inherently difficult. It is no wonder that courts or arbitrators are often asked to distribute equitably a risk to parties who do not perceive the same risks and do not want to assume a disproportionate share of such risks.

DESIGN OF A TIE-BACK RETAINING WALL

This example describes the use of a tie-back retaining wall built in the 1960's when such construction was uncommon and posed a considerable risk. The engineer designing it and the owner were aware of the risk because of potentially extreme financial losses from both remedial and litigation costs in the event that the retaining wall failed and permitted a failure of the slope. But the benefits were perceived as being worth the risk—benefits to the owner in terms of both lower cost and shorter schedule, and benefits to the engineer in terms of professional satisfaction in meeting the owner's needs and solving what appeared to be an insurmountable technical problem.

The tie-back retaining wall was designed to permit a cut in a hillside to provide additional space for the expansion of a steel-making facility. Figure 4.9 shows a cross section of the original hillside located in an urban area. Numerous residential dwellings were located on top of the hill which would have been prohibitively costly or perhaps impossible to remove to permit regrading of the hillside to push back the toe of the slope. The only realistic way of accomplishing the desired goal was to attempt to remove the toe of the existing slope and use a tie-back retaining wall to stabilize the slope as shown in Figure 4.9.

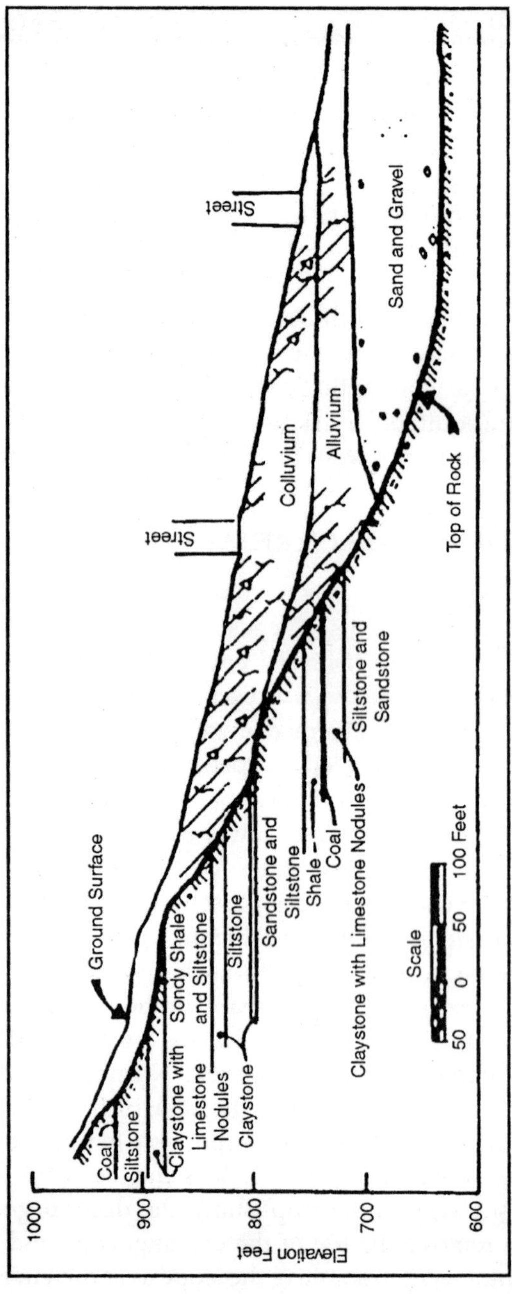

Fig. 4.9 Typical Cross Section of Hillside Adjoining Site

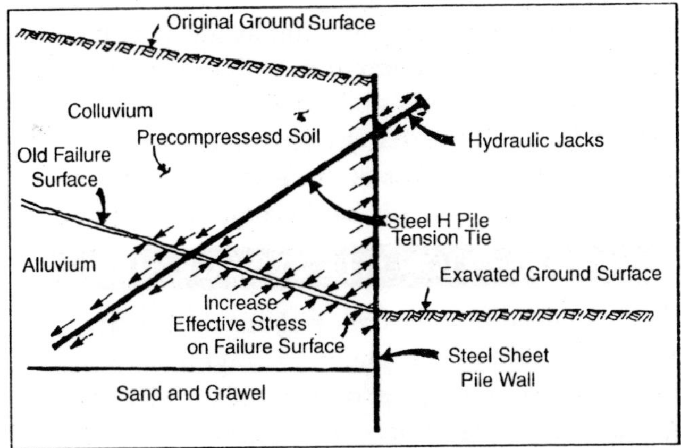

Fig. 4.10 Schematic Section of Anchored Steel Sheet Pile Retaining Wall

A commitment was made by both the owner and the engineer to accomplish what was a common goal. The engineer made a commitment to design and construct the wall in a manner which permitted a real-time evaluation of problems and the ability to take mitigating measures throughout the construction of the wall. The owner made a commitment to give the engineer both the professional latitude and resources required to perform his work. A design-construct contract was negotiated whereby the design could be modified as actual conditions were encountered during construction. But even with all of the planning, investigation and design efforts, there still remained a sizable risk of failure.

The wall was successfully built—not according to a pre-devised plan which went smoothly, and not without numerous problems to be resolved as unexpected groundwater and geological conditions were encountered. Estimated costs were exceeded as each unexpected condition was addressed. But there were no construction delays and their attendant costs as disputes over changed conditions and contract terms were reconciled. There were no costs for legal fees arising from litigation nor increased interest costs as construction stopped while disputes were litigated. The owner paid more than was estimated, but not more than was necessary and not as much as if he had to acquire the property at the top of the hill to regrade the slope. In addition, the owner was able to attain the desired facility expansion in far less time than by any other method.

As a result of the success of this experience and others, the use of tie-back retaining walls has become a routine practice.

CONSTRUCTION SITE ENVIRONMENT

While the general information about the construction site is usually available at the planning stage of a project, it is important for the design professionals and construction manager as well as the contractor to visit the site. Each group will be benefited by first-hand knowledge acquired in the field.

For design professionals, an examination of the topography may focus their attention to the layout of a facility on the site for maximum use of space in compliance with various regulatory restrictions. In the case of industrial plants, the production or processing design and operation often dictate the site layout. A poor layout can cause construction problems such as inadequate space for staging, limited access for materials and personnel, and restrictions on the use of certain construction methods. Thus, design and construction inputs are important in the layout of a facility.

The construction manager and the contractor must visit the site to gain some insight in preparing or evaluating the bid package for the project. They can verify access roads and water, electrical and other service utilities in the immediate vicinity, with the view of finding suitable locations for erecting temporary facilities and the field office. They can also observe any interferences of existing facilities with construction and develop a plan for site security during construction.

In examining site conditions, particular attention must be paid to environmental factors such as drainage, groundwater and the possibility of floods. Of particular concern is the possible presence of hazardous waste materials from previous uses. Cleaning up or controlling hazardous wastes can be extremely expensive.

GROUNDWATER POLLUTION FROM A LANDFILL

The presence of waste deposits on a potential construction site can have substantial impacts on the surrounding area. Under existing environmental regulations in the United States, the responsibility for cleaning up or otherwise controlling wastes generally resides with the owner of a facility in conjunction with any outstanding insurance coverage.

A typical example of a waste problem is illustrated in Figure 4.11. In this figure 4.11, a small pushover burning dump was located in a depression on a slope. The landfill consisted of general refuse and was covered by a very sandy material. The inevitable infiltration of water from the surface or from the groundwater into the landfill will result in vertical or horizontal percolation of leachable ions and organic contamination. This leachate would be odorous and potentially hazardous in water. The pollutant would show up as seepage downhill, as pollution in surface streams, or as pollution entering the regional groundwater.

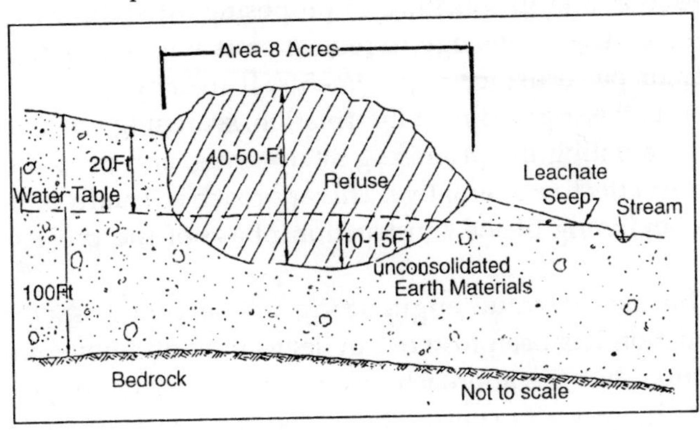

Fig. 4.11 Cross-Section Illustration of a Landfill

Before new construction could proceed, this landfill site would have to be controlled or removed.

Typical control methods might involve:

- Surface water control measures, such as contour grading or surface sealing.
- Passive groundwater control techniques such as

underground barriers between the groundwater and the landfill.

- Plume management procedures such as pumping water from surrounding wells.
- Chemical immobilization techniques such as applying surface seals or chemical injections.
- Excavation and reburial of the landfill requiring the availability of an engineered and environmentally sound landfill.

The excavation and reburial of even a small landfill site can be very expensive. For example, the estimated reburial cost for a landfill like that shown in Figure 4.11 was in excess of $ 4 million in 1978.

VALUE ENGINEERING

Value engineering may be broadly defined as an organized approach in identifying unnecessary costs in design and construction and in soliciting or proposing alternative design or construction technology to reduce costs without sacrificing quality or performance requirements. It usually involves the steps of gathering pertinent information, searching for creative ideas, evaluating the promising alternatives, and proposing a more cost effective alternative. This approach is usually applied at the beginning of the construction phase of the project life cycle.

The use of value engineering in the public sector of construction has been fostered by legislation and government regulation, but the approach has not been widely adopted in the private sector of construction. One explanation may lie in the difference in practice of engineering design services in the public and private sectors. In the public sector, the fee for design services is tightly monitored against the "market price," or may even be based on the lowest bid for service.

Such a practice in setting professional fees encourages the design professionals to adopt known and tried designs and construction technologies without giving much thought to alternatives that are innovative but risky. Contractors are willing to examine such alternatives when offered incentives for sharing the savings by owners. In the private sector, the owner has the freedom to offer such incentives to design professionals as well as the contractors without being concerned about the appearance of favoritism in engaging professional services.

Another source of cost savings from value engineering is the ability of contractors to take advantage of proprietary or unusual techniques and knowledge specific to the contractor's firm. For example, a contractor may have much more experience with a particular method of tunneling that is not specified in the original design and, because of this experience, the alternative method may be less expensive. In advance of a bidding competition, a design professional does not know which contractor will undertake the construction of a facility. Once a particular contractor is chosen, then modifications to the construction technology or design may take advantage of peculiar advantages of the contractor's organization.

As a final source of savings in value engineering, the contractor may offer genuine new design or construction insights which have escaped the attention of the design professional even if the latter is not restrained by the fee structure to explore more alternatives. If the expertise of the contractor can be utilized, of course, the best time to employ it is during the planning and design phase of the project life cycle. That is why professional construction management or integrated design/construction are often preferred by private owners.

CONSTRUCTION PLANNING

The development of a construction plan is very much analogous to the development of a good facility design. The planner must weigh the costs and reliability of different options while at the same time insuring technical feasibility. Construction planning is more difficult in some ways since the building process is dynamic as the site and the physical facility change over time as construction proceeds.

Forming a good construction plan is an exceptionally challenging problem. There are numerous possible plans available for any given project. While past experience is a good guide to construction planning, each project is likely to have special problems or opportunities that may require considerable ingenuity and creativity to overcome or exploit. Unfortunately, it is quite difficult to provide direct guidance concerning general procedures or strategies to form good plans in all circumstances.

There are some recommendations or issues that can be addressed to describe the *characteristics* of good plans, but this does not necessarily tell a planner how to discover a good plan. However, as in the design process, strategies of *decomposition* in which planning is divided into subproblems and *hierarchical planning* in which general activities are repeatably subdivided into more specific tasks can be readily adopted in many cases.

From the standpoint of *construction contractors* or the construction divisions of large firms, the planning process for construction projects consists of three stages that take place between the moment in which a planner starts the plan for the construction of a facility to the moment in which the evaluation of the final output of the construction process is finished.

The *estimate* stage involves the development of a cost and duration estimate for the construction of a facility as part of the proposal of a contractor to an owner. It is the stage in which assumptions of resource commitment to the necessary activities to build the facility are made by a planner. A careful and thorough analysis of different conditions imposed by the construction project design and by site characteristics are taken into consideration to determine the best estimate.

The success of a contractor depends upon this estimate, not only to obtain a job but also to construct the facility with the highest profit. The planner has to look for the time-cost combination that will allow the contractor to be successful in his commitment. The result of a high estimate would be to lose the job, and the result of a low estimate could be to win the job, but to lose money in the construction process. When changes are done, they should improve the estimate, taking into account not only present effects, but also future outcomes of succeeding activities. It is very seldom the case in which the output of the construction process exactly echoes the estimate offered to the owner. In the *monitoring and control stage* of the construction process, the construction manager has to keep constant track of both activities' durations and ongoing costs. It is misleading to think that if the construction of the facility is on schedule or ahead of schedule, the cost will also be on the estimate or below the estimate, especially if several changes are made. Constant evaluation is necessary until the construction of the facility is complete. When work is finished in the construction process, and information about it is provided to the planner, the third stage of the planning process can begin.

The *evaluation* stage is the one in which results of the construction process are matched against the estimate. A planner deals with this uncertainty during the estimate stage. Only when the outcome of the construction process is known is he/she able to evaluate the validity of the estimate. It is in this last stage of the planning process that he or she determines if the assumptions were correct. If they were not or if new constraints emerge, he/she should introduce corresponding adjustments in future planning.

INDUSTRIALIZED CONSTRUCTION AND PRE-FABRICATION

Another approach to construction innovation is to apply the principles and organizational solutions adopted for manufacturing. Industrialized construction and pre-fabrication would involve transferring a significant portion of construction operations from the construction site to more or less remote sites where individual components of buildings and structures are produced. Elements of facilities could be prefabricated off the erection site and assembled by cranes and other lifting machinery. There are a wide variety and degrees of introducing greater industrialization to the construction process. Many components of constructed facilities have always been manufactured, such as air conditioning units. Lumber, piping and other individual components are manufactured to standard sizes. Even temporary items such as forms for concrete can be assembled off-site and transported for use. Reinforcing bars for concrete can also be pre-cut and shaped to the desired configuration in a manufacturing plant or in an automated plant located proximate to a construction site. A major problem in extending the use of pre-fabricated units is the lack of standardization for systems and building regulations. While designers have long adopted standard sizes for individual components in designs, the adoption of standardized sub-assemblies is rarer. Without standardization, the achievement of a large market and scale economies of production in manufacturing may be impossible. An innovative and more thorough industrialization of the entire building process may be a primary source of construction cost savings in the future.

PLANNING OF PRE-FABRICATION

When might pre-fabricated components be used in preference to components assembled on a construction site? A straightforward answer is to use pre-fabricated components whenever their cost, including transportation, is less than the cost of assembly on site. As an example, forms for concrete panels might be transported to a construction site with reinforcing bars already built in, necessary coatings applied to the forms, and even special features such as electrical conduit already installed in the form. In some cases, it might be less expensive to pre-fabricate and transport the entire concrete panel to a manufacturing site. In contrast, traditional construction practice would be to assemble all the different features of the panel on-site. The relevant costs of these alternatives could be assessed during construction planning to determine the lowest cost alternative.

In addition to the consideration of direct costs, a construction planner should also consider some other aspects of this technology choice. First, the planner must insure that pre-fabricated components will satisfy the *relevant building codes* and *regulations*. Second, the *relative quality* of traditional versus pre-fabricated components as experienced in the final facility should be considered. Finally, the *availability of components* at the required time during the construction process should also be considered.

IMPACTS OF BUILDING CODE

Building codes originated as a part of the building regulatory process for the safety and general welfare of the public. The source of all authority to enact building codes is based on the police power of the state which may be delegated by the state legislature to local government units. Consequently, about 8,000 localities having their own building codes, either by following a national model code or developing a local code.

The lack of uniformity of building codes may be attributed to a variety of reasons:

- Neighboring municipalities may adopt different national models as the basis for local regulation.
- Periodic revisions of national codes may not be adopted by local authorities before the lapse of several years.
- Municipalities may explicitly decline to adopt specific provisions of national model codes or may use their own variants of key provisions.
- Local authorities may differ in interpretation of the same language in national model codes.

The lack of uniformity in building codes has serious impact on design and construction as well as the regulatory process for buildings.

Among the significant factors are:

- Delay in the diffusion of new building innovations which may take a long time to find their ways to be incorporated in building codes.
- Discouragement to new production organizations, such as industrialized construction and prefabrication.
- Duplication of administrative cost of public agencies and compliance cost incurred by private firms.

COMPUTER-AIDED ENGINEERING

In the past twenty years, the computer has become an

essential tool in engineering, design, and accounting. The innovative designs of complicated facilities cited in the previous sections would be impossible without the aid of computer based analysis tools. By using general purpose analysis programmes to test alternative designs of complex structures such as petrochemical plants, engineers are able to greatly improve initial designs. General purpose accounting systems are also available and adopted in organizations to perform routine bookkeeping and financial accounting chores. These applications exploit the capability for computers to perform numerical calculations in a pre-programmed fashion rapidly, inexpensively and accurately.

Despite these advances, the computer is often used as only an incidental tool in the design, construction and project management processes. However, new capabilities, systems and application programmes are rapidly being adopted. These are motivated in part by the remarkable improvement in computer hardware capability, the introduction of the Internet, and an extraordinary decline in cost. New concepts in computer design and in software are also contributing. For example, the introduction of personal computers using micro circuitry has encouraged the adoption of interactive programmes because of the low cost and considerable capability of the computer hardware. Personal computers available for a thousand dollars in 1995 have essentially the same capability as expensive mainframe computer systems of fifteen years earlier.

Computer graphics provide another pertinent example of a potentially revolutionary mechanism for design and communication. Graphical representations of both the physical and work activities on projects have been essential tools in the construction industry for decades. However, manual drafting of blueprints, plans and other diagrams is laborious and expensive. Stand alone, computer aided drafting equipment has proved to be less expensive and fully capable of producing the requiring drawings. More significantly, the geometric information required for producing desired drawings might also be used as a database for computer aided design and computer integrated construction. Components of facilities can be represented as three dimensional computer based *solid models* for this purpose. Geometric information forms only one component of integrated design databases in which the computer can assure consistency, completeness and compliance with relevant specifications and constraints. Several approaches to integrated computer aided engineering environments of this type have already been attempted.

Computers are also being applied more and more extensively to non-analytical and non-numerical tasks. For example, computer based specification writing assistants are used to rapidly assemble sets of standard specifications or to insert special clauses in the documentation of facility designs. As another example, computerized transfer of information provides a means to avoid laborious and error-prone transcription of project information.

While most of the traditional applications and research in computer aids have emphasized numerical calculations, the use of computers will rapidly shift towards the more prevalent and difficult problems of planning, communication, design and management.

Knowledge based systems represent a prominent example of new software approaches applicable to project management. These systems originally emerged from research in artificial intelligence in which human cognitive processes were modeled. In limited problem domains such as equipment configuration or process control, knowledge based systems have been demonstrated to approach or surpass the performance of human experts. The programmes are marked by a separation between the reasoning or "inference" engine programme and the representation of domain specific knowledge. As a result, system developers need not specify complete problem solving strategies (or algorithms) for particular problems. This characteristic of knowledge based systems make them particularly useful in the ill-structured domains of design and project management. Chapter 15 will discuss knowledge based systems in greater detail.

Computer programme assistants will soon become ubiquitous in virtually all project management organizations. The challenge for managers is to use the new tools in an effective fashion. Computer intensive work environments should be structured to aid and to amplify the capabilities of managers rather than to divert attention from real problems such as worker motivation.

PRE-PROJECT PLANNING

Even before design and construction processes begin, there is a stage of "pre-project planning" that can be critical for project success. In this process, the project scope is established. Since construction and design professionals are often not involved in this project scope stage, the terminology of describing this as a "pre-project" process has arisen. From the owner's perspective, defining the project scope is just another phase in the process of acquiring a constructed facility.

The definition of a project scope typically involves developing project alternatives at a conceptual level, analyzing project risks and economic payoff, developing a financial plan, making a decision to proceed (or not), and deciding upon the project organization and control plan. The next few chapters will examine these different problems at some length.

The danger of poor project definition comes from escalating costs (as new items are added) or, in the extreme, project failure. A good definition of scope allows all the parties in the project to understand what is needed and to work towards meeting those needs.

PROJECT DEFINITION RATING INDEX (PDRI)

The Construction Industry Institute has developed rating indexes for different types of projects to assess the adequacy of project scope definitions. These are intended to reflect best practices in the building industry and provides a checklist for recommended activities and milestones to define a project scope. The rating index is a weighted sum of scores received for a variety of items on the scope definition checklist. Each item in the checklist is rated as "not applicable" (0), "complete definition" (1), "minor deficiencies" (2), "some deficiencies" (3), "major deficiencies" (4) or "incomplete or poor definition" (5). *Lower scores in these categories are preferable. Some items in the checklist include*:

- Business Strategy for building use, justification, plan, economic analysis, facility requirements, expansion/ alteration consideration, site selection issues and project objectives.
- Owner Philosophy with regard to reliability, maintenance, operation and design.
- Project Requirements for value engineering, design, existing facility, scope of work review, schedule and budget.
- Site Information including applicable regulatory reporting and permits requirements.
- Building Programming including room by room definitions for use, finishes, interior requirements and hvac (heating, ventilating and air conditioning).
- Design Parameters including all components and a constructability analysis.
- Equipment including inventory, locations and utility requirements.

5

Site Works

SITE SURVEY

Construction is one of the largest industries in the world. Surveying plays an extremely important role in any construction project. Construction surveying can take many forms. It is used to establish the location and alignment of highways, bridges, buildings, pipes, and other man-made objects. After large-scale projects are completed, an "as-built" survey is performed to locate any modifications that were made to the plans during construction.

Highway surveys involve the location of alignments and computation of volumes materials that must be added, removed, or moved. It initially requires a topographic survey of the site. For large projects, photogrammetric methods will be used to develop the *base* map. The base map is used by surveyors and other professional to create a base plan for the project. After the alignment has been established, the quantities of earth that must be added or removed are computed. The goal of most projects is to minimize the hauling distances of the earth. This is done using *mass diagrams*. Eventually surveyors layout the elevation and slope of the various subgrades, base, and top coat materials. The end result is a smooth alignment with smooth transitions from straight to curved sections allowing for safe public transportation.

Fig 5.1

Traditionally, a highway alignment survey involves the placement of wooden stakes to mark the highway's location. Today, GPS equipment is being used to provide real-time positioning and alignment of construction equipment. This form of *stake less* construction has changed the requirements of surveying personnel working in construction. In addition to the computations required for placement of alignments, today's construction surveyor needs a strong background in coordinate systems, map projections, and geodetic principles.

Fig 5.2

Building construction is another area of construction surveying. In this field, the correct placement of footings, foundations, piers and other items of building construction are essential for a sound structure. A university education can provide you with the knowledge required to practice in the area construction surveying.

Fig 5.3

ROUTE SURVEYS

A route survey, as the name implies, is a survey that deals with the route or course that a highway, road, or utility line will follow.

While the end product of a route survey for a highway certainly differs from that for a utility line, it may, nevertheless, be said that the purposes of any route survey are to:

- Select one or more tentative general routes for the roadway or utility,
- Gather enough information about the general route to make it possible for designers to select the final location of the route, and
- Mark this final location.

Consistent with these purposes, a route survey is usually broken down into reconnaissance, preliminary, and final-location survey phases that satisfy, respectively, each of the purposes given above. Sometimes, however, circumstances may preclude the requirement to perform all three phases; for example, if a new road or utility line is to be constructed on a military installation having well-marked vertical and horizontal control networks and up-to-date topographic maps and utility maps, then perhaps the reconnaissance and preliminary survey phases would not be required.

Chapter of the EA3 TRAMAN discusses each phase of route surveying as applied to roads and highways. That discussion is presented in sufficient enough depth to preclude the need to further discuss highway route surveying in this TRAMAN. You should, however, review that discussion and read other publications dealing with the subject of route surveying. Aside from roads and highways, other uses of route surveys are for aboveground utility lines-most commonly power and communication lines—and for underground utilities, such as power, communication, sewer, water, gas, and fuel lines. The character of the route survey for a utility will vary, of course, with different circumstances; for example, a sanitary sewer, water distribution line, or an electrical distribution line in an urban area will generally follow the streets on which the buildings it serves are located.

Also, since these areas will, in all likelihood, have other existing utilities, there should be existing utilities maps that can be used in the design of the new utility line. Consequently, in cases such as this, reconnaissance and preliminary surveys are seldom necessary. On the other hand, a power transmission line or other utility running through open country on a large military installation may require reconnaissance and preliminary surveys in addition to the final-location survey.

Route Surveys for Overhead Electrical Distribution and Transmission Lines

The reconnaissance survey for electrical power lines employs many of the same principles and practices that you studied for highway work; however, the design considerations are different. For a power line, the design engineer considers principles that you studied in chapter 2 of this TRAMAN to select one or more tentative routes over which the line will pass.

For convenience, those principles are listed as follows:

- Select the shortest possible route.
- Follow the highways and roads as much as possible.
- Follow the farmer's property or section lines.
- Route in the direction of possible future loads.
- Avoid going over hills, ridges, swamps, and bottom lands.
- Avoid disrupting the environment.

During the reconnaissance phase, you should first study all available maps of the area to gain a general understanding of the landscape. If a portion of the line is off the military installation, determine the owner-ship of the lands through which the line will pass. That is necessary to obtain permission to run the line.

Look for any existing utilities that may already exist in the area. If there are existing utilities, then look for existing utilities maps. Visit the area to examine the terrain and look for any natural or man-made features that may hinder or help the construction. In short, gather all information that the engineer will need to select one or more general routes for the power line.

With the tentative route or routes selected, you are ready to conduct a preliminary survey from which a map is prepared showing the country over which the line will pass. Since the final location is not known, a wide strip of land needs to be mapped. When running the preliminary survey, incorporate all pertinent topographic information into the field notes. Note particularly any existing overhead or underground lines and indicate whether they are power or communications lines. Locate such features as hills, ridges, marshes, streams, forests, roads, railways, power plants, buildings, and adjacent military camps or bases.

When the preliminary mapping is completed, the engineer selects the final route. Again, the engineer considers the principles listed above to select theroute.

POLE LINE SURVEYS

When the route has been selected, a plan and profile are plotted. The plan shows the route the line will follow and the significant topography adjacent to the route. The profile shows the ground elevation along the line and the top eleva-tions of the poles. These elevations are set in accord-ance with minimum allowable clearances specified in the *National Electrical Safety Code* (NESC), ANSI C2, and the most recent edition of the *National Electrical*

For distribution lines, poles should be placed on the side of the street that is most free of other lines and trees. Try to keep off the main streets. As much as possible, you should use the same side of the road throughout the length of the line. For straight portions of lines, the usual spacing between poles is about 125 feet (100 feet minimum and 150 feet maximum); however, to make the poles come in line with property lines or fences, the span length may need to be adjusted. The engineer will determine the spans. Along roads, poles should be placed 2 feet from the inside edge of the curb or 2 feet from the edge of the road surface where curbs do not exist. On open roadways or highways, poles should be set 18 inches from the outside of fences.

For transmission lines, poles should be located in high places so that shorter poles can be used and still maintain the proper ground clearance at the middle of the span. Avoid locating poles along the edge of embankments or streams where washouts can be expected.

In rolling country, the grading of the line should be considered when determining pole locations. A well-graded line does not have any abrupt changes up or down the line and will appear nearly horizontal regardless of small changes in ground level. Sometimes, by shifting a pole location a few feet, a standard length pole can be used where otherwise an odd-sized pole would be needed. In addition, transmission line poles should be located at least 2 feet from curbs, 3 feet from fire hydrants, 12 feet from the nearest track of a railroad track, and 7 feet from railway sidings.

When you are staking pole locations, the center of each pole is marked with a hub on the line; the hub may be offset. On the guard stake, you put the pole number, the line elevation, and the distance from the top of the hub to the top of the pole obtained from the profile.

TOWER LINE SURVEYS

High-voltage lines are often supported by broad-based steel towers. For a tower line, construction economy requires that changes in direction be kept at a minimum. That is because a tower located where a line changes direction must withstand a higher stress than one located in a straight direction part of the line. In general, tower construction is cheaper in level country than in broken country; however, the line may be run over broken country to minimize changes in direction, to make the distance shorter, or to follow a line where the cost of obtaining right-of-way is inexpensive.

Lines should be located adjacent to existing roads, whenever practical, to provide easier access for construction and future maintenance. When a change in direction in a tower line is unavoidable, it should be made gradually in as small-angular increments as possible. Suppose, for example, a change in direction of 90° is required. Instead of an abrupt change in direction of 90°, towers should be set so as to cause the line to follow a gradual curve in a succession of chords around an arc of 90°.

Route Surveys for Drainage

When man-made structures are erected in a certain area, it is necessary to plan, design, and construct an adequate drainage system. Generally, an underground drainage system is the most desirable way to remove surface water effectively from operating areas. An open drainage system, like a ditch, is economical; however, when not properly maintained, it is unsightly and unsafe. Sometimes, an open drainage system also causes erosion, thus resulting in failures to nearby structures. Flooding caused by an inadequate drainage system is the most prevalent cause leading to the rapid deterioration of roads and airfields.

DRAINAGE SYSTEM

Sanitary sewers carry waste from buildings to points of disposal; storm sewers carry surface run-off water to natural watercourses or basins. In either case the utility line must have a gradient; that is, a downward slope towards the disposal point, just steep enough to ensure a gravity flow of waste and water through the pipes. This gradient is supplied by the designing engineer.

Natural Drainage

To understand the con-trolling considerations affecting the location and other design features of a storm sewer, you must know something about the mechanics of water drainage from the earth's surface.

When rainwater falls on the earth's surface, some of the water is absorbed into the ground. The amount absorbed will vary, of course, according to the physical characteristics of the surface. In sandy soil, for instance, a large amount will be absorbed; on a concrete surface, absorption will be negligible. Of the water not absorbed into the ground, some evaporates, and some, absorbed through the roots and exuded onto the leaves of plants, dissipates through a process called transpiration.

The water that remains after absorption, evaporation, and transpiration is technically known as run-off. This term relates to the fact that this water, under the influence of gravity, makes its way (that is, runs off) through natural channels to the lowest point it can attain.

To put this in terms of a general scientific principle, water, whenever it can, seeks its own level. The general, final level that unimpeded water on the earth's surface seeks is sea level; and the rivers of the earth, most of which empty into the sea, are the earth's principal drainage channels. However, not all of the earth's run-off reaches the great oceans; some of it is caught in landlocked lakes, ponds, and other nonflowing inland bodies of water.

Let's consider, now, a point high in the mountains somewhere. As rain falls in the area around this point, the run-off runs down the slopes of a small gully and forms a small stream, which finds a channel down-ward through the ravine between two ridges.

As the stream proceeds on its course, it picks up more and more water draining in similar fashion from high points in the area through which the stream is passing. As a result of this continuing accumulation of run-off, the stream becomes larger until eventually it either becomes or joins a large river making its way to the sea—or it may finally empty into a lake or some other inland body of water. In normal weather conditions, the natural channels through which this run-off passes can generally contain and dispose of all the run-off. However, during the winter in the high mountains, run-off is commonly interrupted by snow conditions; that is, instead of running off, the potential run-off accumulates in the form of snow. When this accumulated mass melts in the spring, the run-off often attains proportions that overwhelm the natural channels, causing flooding of surrounding areas. In the same fashion, unusually heavy rainfall may overtax the natural channels.

Artificial Drainage

When artificial structures are introduced into an area, the natural drainage arrangements of the area are upset. When, for example, an area originally containing many hills and ridges is graded off flat, the previously existing natural drainage channels are removed, and much of the effect of gravity on run-off is lost. When an area of natural soil is covered by artificial paving, a quantity of water that previously could have been absorbed will now present drainage problems.

In short, when man-made structures, such as bridges, buildings, and so forth, are erected in an area, it is usually necessary to design and construct an artificial drainage system to offset the extent to which the natural drainage system has been upset. Storm sewers are usually the primary feature of an artificial drainage system; however, there are other features, such as drainage ditches.

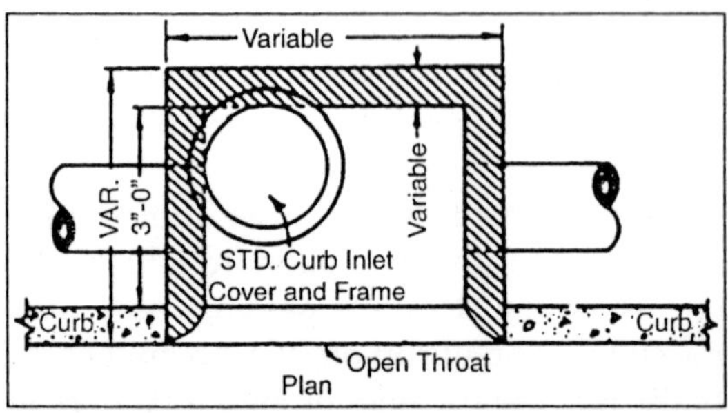

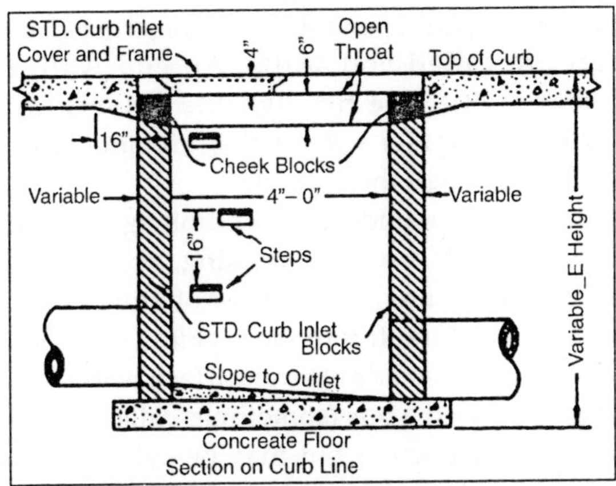

Fig. 5.4 Working Drawing for a Typical Curb Inlet.

Both storm sewers and ditches carry surface run-off. The only real difference between a drainage ditch and a storm sewer is the fact that the ditch lies on the surface and the storm sewer lies below the surface.

Similarly, there is no essential difference in mechanical principle between an artificial and a natural drainage system. Like a natural channel, an artificial channel must slope downward and must become progressively larger as it proceeds along its course, picking up more run-off as it goes. Like a natural system, an artificial system must reach a disposal point—usually a stream whose ultimate destination is the sea or a standing inland body of water. At the terminal point of the system where the accumulated run-off discharges into the disposal point, the run-off itself is technically known as discharge. The discharge point in the system is called the outfall.

Ditches

A surface drainage system consists principally of ditches that form the drainage channels. A ditch may consist simply of a depression formed in the natural soil, or it may be a paved ditch. Where a ditch must pass under a structure (such as a highway embankment, for example), an opening called a culvert is constructed. A pipe culvert has a circular opening; a box culvert has a rectangular opening. Walls constructed at the ends of a culvert are called end walls. An end wall, running perpendicular to the line through the culvert, may have extensions called wings (or wing walls), running at an oblique angle to the line through the culvert.

Storm Sewers.—An underground drainage system (that is, a storm sewer) consists, broadly speaking, of a buried pipeline called the trunk or main, and a series of storm water inlets, which admit surface run-off into the pipeline. An inlet consists of a surface opening that admits the surface water run-off and an inner chamber called a box (sometimes called a catch basin). A box is usually rectangular but may be cylindrical. An inlet with a surface opening in the side of a curb is called a curb inlet. A working drawing of a curb inlet is shown in figure 5.5. An inlet with a horizontal surface opening covered by a grating is called a grate (sometimes a drop) inlet. A general term applied in some areas to an inlet that is neither a curb nor a grate inlet is yard inlet.

Appurtenances

Technically speaking, the term *storm sewer* applies to the pipeline; the inlets are called appurtenances. There are other appurtenances, the most common of which are manholes and junction boxes. A manhole is a box that is installed, of necessity, at a point where the trunk changes direc-tion, gradient, or both. The term *manhole* originally related to the access opening at one of these points; however, a curb inlet and a junction box nearly always have a similar access opening for cleaning, inspection, and maintenance purposes. One of these openings is often called a manhole, regardless of where it is located. However, strictly speaking, the access opening on a curb inlet should be called a curb-inlet opening; and on a junction box, a junction-box opening. Distances between manholes are normally 300 feet, but this distance may be extended to a maximum of 500 feet when specified. The access opening for a manhole, curb inlet, or junction box consists of the cover and a supporting metal frame. A frame for a circular cover is shown in figure 5.5. Some covers are rectangular. The frame usually rests on one or more courses of adjusting blocks so that the rim elevation of the cover can be varied slightly to fit the surface grade elevation by varying the vertical dimensions, or the number of courses, of the adjusting blocks. A junction box is similar to a manhole but is installed, of necessity, at a point where two or more trunk lines converge. The walls of an inlet, manhole, or junction box maybe constructed of special concrete masonry units or of cast-in-place concrete.

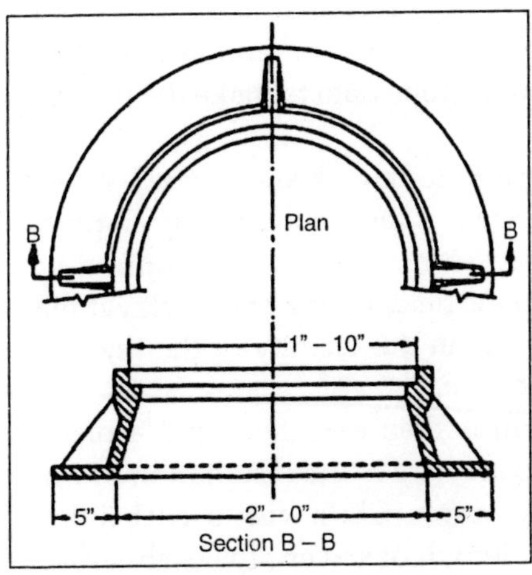

Fig. 5.5 Frame for an Access Opening

The bottom consists of a formed slab, sloped in the direction of the line gradient and often shaped with channels for carrying the water across the box from the inflowing pipe to the out flowing pipe.

STORM SEWER ROUTE SURVEY

The character of the route survey for a storm sewer depends on the circumstances. The nature of the ground may be such as to indicate, without the necessity for reconnaissance and preliminary location surveys, just where the line must go. This is likely to be the case in a development area; that is, an area that will be closely built up and in which the lines of the streets and locations of the buildings have already been determined. In these circumstances, the reconnaissance and preliminary surveys may be said to be done on paper.

On the other hand, a line—or parts of it—often must be run for considerable distances over rough, irregular country. In these circumstances the route survey consists of reconnaissance, preliminary location, and final-location surveys. If topographic maps of the area exist, they are studied to determine the general area along which the line will be run. If no such maps exist, a reconnaissance party must select one or more feasible route areas, run random traverses through these, and collect enough tope data to make the planning of a tentative route possible.

After these data have been studied, a tentative route for the line is selected. A preliminary survey party runs this line, making any necessary adjustments required by circumstances encountered in the field, taking profile elevations, and gathering enough tope data in the vicinity of the line to make design of the system possible.

The system is then designed, and a plan and profile are made. The project here is the installation of 230 feet of 18-inch concrete sewer pipe (CSP) with a curb inlet (CI "A"). The computational length of sewer pipe is always given in terms of horizontal feet covered. The actual length of a section is, of course, greater than the computational length because of the slope. The pipe is to run down slope from a curb inlet to a manhole in an existing sewer line. The reason for the distorted appearance of the curb inlet and manhole, which look much narrower than they would in their true proportions, is the exaggerated vertical scale of the profile. The appearance of the pipe is similarly distorted.

The pipe to be installed is to be placed at a gradient of 2.39 per cent. The invert elevation of the out flowing 21-inch pipe at the manhole is 91.47 feet; that of the inflowing 18-inch pipe is to be 92.33 feet. Obviously, there is a drop here of 0.86 foot. Of this drop, 0.25 foot is because of the difference in diameters; the other 0.61 foot is probably because of structural and velocity head losses. From the invert in at the manhole, the new pipe will extend 230 horizontal feet to the invert at the center line of the curb inlet. The difference in elevation between the invert elevation at the manhole and the invert elevation at the curb inlet will be the product of 2.39 (the grade percentage) times 2.30 (number of 100-foot stations in 230 horizontal feet), or 5.50 feet. Therefore, the invert elevation at the curb inlet will be 92.33 feet (invert elevation at the manhole) plus 5.50 feet, or 97.83 feet. The invert elevation at any intermediate point along the line can be obtained by similar computation.

The plan shown in figure 5.6 is greatly simplified for the sake of clearness—it contains the bare minimum of data required for locating the new line. Plans used in actual practice usually contain more information.

The plan and profile constitute the paper location of the line. A final-location survey party runs the line in the field. Where variations are required because of circumstances discovered in the field (such as the discovery of a large tree or some similar obstruction lying right on the line), the direction of the line is altered (after receiving approval to do so) and the new line is tied to the paper location. The final-location party may simply mark the location of the line and take profile elevations, or it may combine the final-location survey and the stakeout (which is part of the construction survey, rather than the route survey) in the same operation.

Other Route Surveys

While highways and the various types of utilities have differing design requirements that must be considered when conducting route surveys, you haveprobably observed in your studies that much of route surveying is similar regardless of the type of construction being planned. This is especially true during the reconnaissance phase. Therefore, with a firm understanding of the preceding paragraphs and of the EA3 TRAMAN discussion of route surveying, you should have little difficulty in planning and performing other types of route surveys. For roads and highways, however, you also must have an

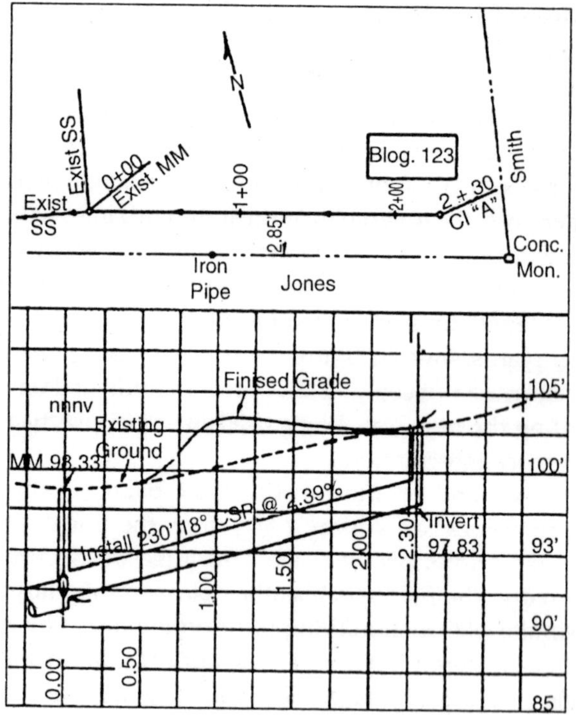

Fig. 5.6 Storm Sewer Plan and Profile

SITE INVESTIGATION

PLANNING A SITE INVESTIGATION

Doubleton and West have discussed the planning and direction of site investigations. They state that "the main investigation is the full investigation of the site using boreholes and trial pits and includes the preparation of the site-investigation report with revised plans and sections, interpretation and recommendations for design."

They consider that there are two aspects to the site investigation. The geological structure and character of the site and the testing of the soil both in the laboratory and in-situ.

They suggest that the planning should consider the following questions:

- Is the succession of strata known over the whole site and is there correlation across the whole site known? What measurements of depth dip and outcrop are required to clarify the stratigraphy? Should any key measurements, such as depth to bedrock or peat, be made?
- Are the different strata fairly homogeneous over the site or do local variations exist? Are there more complex areas of strata that require investigation or closer examination during construction? Will there be areas where the excavated material will be unsuitable for fill and will need to be replaced? Are there areas where needs to be assessed to ascertain working methods?
- Are there ground structures that need closer examination-for example the extent of disturbed strata, the location and extent of natural cavities and mine workings, their liability to cause subsidence or movement, the extent and depth of surface movement and instability? Is there other phenomenon likely to give rise to instability such as fractures and river crossings or alluvial areas that may have buried soft material or peat? Are there likely to be any undetected structures?
- Will any part of the route be subject to flooding? What contact will there be with water bearing strata and will ground water lowering methods be required during construction?
- Do requirements for the carrying out of special in-situ tests or the taking of undisturbed samples affect the conduct of the qualitative investigation? For example, with forethought a single trial pit may be made to serve both for examining ground materials and structure, and for the in-situ testing and the taking of block samples.

Doubleton and West state that these questions should be asked for the whole route, and must be related to the proposed geometry of the road and it's structure. Investigations must be carried out to the depth at which ground conditions cease to affect the work. The more complex the situation, the more extensive the investigation will need to be.

The investigation must then be planned to answer the above questions and associated problems. As much information should be obtained from the points examined as each point is expensive. These are frequently used to clarify the interpretation of the site as a whole. Bridge sites, high embankments and deep cuttings are all points of engineering complexity and should be examined thoroughly.

GROUND INVESTIGATION

Ground investigation is taken to be that other than the information available from the walk over survey as discussed previously.

There are two principal methods of investigating the ground conditions, trial pits and boreholes. In addition, the reader should be aware of geophysical techniques such as seismic surveys, which are not discussed here.

Trial Pits

Trial pits are shallow excavations going down to a depth no greater 6m. The trial pit as such is used extensively at the surface for block sampling and detection of services prior to borehole excavation.

Depth	Excavation Method
0-2m	By Hand
2-4m	Wheeled Back Hoe
4-6m	Hydraulic Excavator

An important safety point to note is that *ALL pits below a depth of 1.2m must be supported*. In addition care should be taken as gases such as methane and carbon dioxide can build up in a trial pit. Breathing apparatus must therefore be used if no gas detection equipment is available.

Support for a trial pit generally takes one of three forms:

1. Timbering
2. Steel frames with hydraulic jacks
3. Battered or tapered sides

Three types of sample can be taken from a trial pit:

1. *Disturbed sample*: Samples where the soils in-situ properties are not retained.
2. *Block sample*: A sample that is not undisturbed but retains some in-situ properties.
3. *Push in tube sample*: Tube samples of the soil in a trial pit.

When preparing a trial pit log, the following information should be included. The location, orientation and size of the pit; sketches of faces; depth scale; root structur; water level; seepage. In addition the weather at the time of sampling should be noted as many soils are weather dependant.

BOREHOLES

A borehole is used to determine the nature of the ground (usually below 6m depth) in a qualitative manner and then recover undisturbed samples for quantitative examination. Where this is not possible, for in gravelly soils below the water table, in-situ testing methods are used.

Obviously the information gained from a borehole is an extremely limited picture of the subsurface structure. It is therefore essential to compare the results obtained with those that could have been expected from the desk study. The greater the number of boreholes the more certain it is possible to be of the correlation and thus to trust in the results. The two principal types of boring machine used for Site Investigation in the United Kingdom are light percussive and drilling machines.

Light Percussive is the process of making boreholes by striking the soil then removing it and the most common method is the shell and auger. This is a general term to describe various tools suspended from a triangular tripod incorporating a power winch. The tools are repeatedly dropped down the borehole while suspended by wire from the power winch.

The different tools used include:

- *Clay Cutter*: Used in cohesive materials and is raised and lowered, using it's own weight to cut into the material.
- *Shell*: Used for boring in silts and sands. Similar to the clay cutter, but has a trap door at the bottom to catch material.
- *Chisel*: Used for breaking up hard material such as boulders or rocks. Additional payment is required for chiseling as per the Bill of Quantities and permission is normally required from the Resident Engineer before work can start.

Drilling is the process of boring normally by using a combination of a rotating action and a hydraulic ram. There are many different types of rig depending on access and type of ground expected. Hollow drilling rods enable a flush of water, air, foam or mud which is used to carry the cuttings to the surface as well as lubricating and cooling the drill bit.

The three main types of drill bit are:

1. *Double tube* is where the outer tube rotates and allows for the removal of the cuttings while the inner tube is stationary and prevents the core from shearing. There are different designs of tube varying the location of the flush discharge so as to prevent sample erosion. It is necessary for the hole to be bigger than the tube and so the diamond bits are attached to the outside of the hole, thus allowing the flush to return to the surface.
2. *Triple tube* in corporate a third tube to protect the core even further during extrusion and can have either a split tube, which is removed, or a plastic tube to provide longer term protection. A less effective alternative is to incorporate a nylon liner in a double tube.
3. *Retractable triple tube* is a variation where the inner tube is attached to a retractor and can extend beyond the cutting edge. This gives complete protection to the core

in softer rock whilst in harder rock where this is not necessary; it retracts to become a standard triple tube. This is used in alternating soft/hard rock, typical of a weathered profile.

Core bits are usually diamond tipped and are either surface set, where diamonds are mounted into a matrix, or impregnated where a fine diamond dust is used in the matrix. In softer rocks, the cuttings can clog up the matrix so the softer the rock, the larger the diamonds need to be. Tungsten carbide bits can also be used in the softer rocks.

SAMPLING

Sampling can be either undisturbed, of which in-situ testing is a form, or disturbed.

The principal sampling methods used in boreholes are:

- *SPT test*: This is a dynamic test as described in BS1377 (Part 9) and is a measure of the density of the soil. The test incorporates a small diameter tube with a cutting shoe known as the 'split barrel sampler' of about 650mm length, 50mm external diameter and 35mm internal diameter. The sampler is forced into the soil dynamically using blows from a 63.5kg hammer dropped through 760mm. The sampler is forced 150mm into the soil then the number of blows required to lower the sampler each 75mm up to a depth of 300mm is recorded. This is known as the "N" value. For coarse gravels the split barrel is replaced by a 60 degree cone.
- *Core Sample*: Core samples must be sealed with parafin to maintain the water conditions and then end sealed to prevent physical interference. The mpst common of these is the U100 although other sizes from 54mm to 100mm diameter are used. The standard U100 has a sample area ratio of 30% so large amounts of soil are displaced. A thin walled *Piston Sampler* reduces this to 10%. The sample is pushed or jacked into the ground as opposed to a dynamic action.

- *U100*: This is a 450mm long, 100mm diameter undisturbed sample. The tube has a cutter at one end and the driving equipment at the other. Behind the cutter is a core catcher, incorporating 3 arms that go into the sample as it is withdrawn, to prevent the sample from falling out. Care should be taken to ensure that the cutting shoe is as clean and sharp as possible.
- *Bulk Samples*: Usually taken from trial pits or in soils where there is little or no cohesion. Often called block samples.
- *Water Samples*: Water samples should be taken as soon as water is first struck and the depth recorded. After a suitable period of time (usually 10-15 mins) the depth should be re-recorded and a further sample taken. A final sample should be taken at the end of the borehole and the depth to water regularly recorded. The sample is taken using a device known as a bailer, made from teflon or plastic it incorporates a float to trap the water and should be cleaned after each sample.

The sampling procedure varies according to the type of strata in which the investigation takes place. A recommended sampling procedure is listed below.

Clays

Normally need undisturbed samples:

- U100 every 1.5m or change of stratum. Blow count and penetration should be noted.
- If unable to obtain a U100 then bulk samples as above.
- If U100 does not full penetrate SPT test is required.

Sands and Gravels

Undisturbed samples are not practical due to the lack of cohesion:

- SPT every 1m or change of stratum. Number of seating blows should also be recorded.
- Bulk samples to be taken between SPT's.

Silts

- Alternate SPT and U100 samples at 0.75m intervals

Rock

- SPT on penetrating rock, every 1m and change of stratum where possible
- In softer rocks (Chalks, Marls) U100 may be possible
- Rock must be penetrated at least 1.5m to ensure it isn't a large cobble
- Obtain permission from Resident Engineer before pulling off-site
- If SPT refusal (>50 blows) record number of blows and penetration

REPORTING

The Site Investigation report for a highway design scheme should answer all the questions set out in the planning phase of the Investigation. This should include an assesment of the viablilty of the proposed route and indication of any alternatives.

Included in the report should be a location of all the boreholes, trial pits, other excavations and their logs. These logs should give as much information as possible on the soil and rock structure as it is possible to obtain.

The soil and rock descriptions should be as defined in BS5930 and should contain the information described below:

Soil Description

Often remebered using the acronym MCCSSOW obviously!

- *Moisture Content*: Dry, slightly moist, moist, very moist or wet. Not the measured value just the way it appears in the hand.
- *Colour*: This is an indicator of chemical and mineralogical content. Charts are available but not often used.
- *Consistency*: Loose or dense and other descriptions dependant on soil type. An approximate relationship

can be made between stiffness and untrained shear strength (Cu) and between density and the SPT 'N' Values.

	Cu		SPT 'N' Value
Very Soft	<20	Very Loose	<4
Soft	20-40	Loose	4-10
Firm	40-75	Medium dense	10-30
Stiff	75-150	Dense	30-50
Very stiff	>150	Very dense	>50

- *Structure*: Bedding laminates fissure, joints, fractures, shear zones etc.
- *Soil Type*: Given by particle sizes as described in BS5930 Table
- *Origin*: Try and identify geological area and stereographic unit. This is difficult and often impossible
- *Groundwater Conditions*: Depth to groundwater and any other observations.

Rock Descriptions

The acronym makers came up with CGTSWROS in a moment of inspiration:

- *Colour*: Same terminology as for soils with principal and secondary
- *Grain size*: Range of sizes present and the dominant sizes.
- *Texture and Fabric*: Porphyritic, crystalline, granular, glassy, amorphous, homogeneous and many more as described in BS5930 Table.
- *Structure*: Dependant on the type of rock, reference should be made to BS5930 Table. Discontinuities in the rock can be caused by the drilling action, weathered surfaces indicate natural and clean surface indicate recent fractures.
- *Weathering*: Engineers grade from 1-6 with 1 being fresh and 6 being residual soil with all the rock converted to soil.

- *Rock type*: Reference should be made to BS5930 Table.
- Other Stereographic information, geological period, presence of fossils or coral seams.
- *Strength*: Defined as below from field observations.

Term	Field recognition	Cu (MPa)	Point Load Strength (MPa)
Extremely Strong	Rocks Ring on hammer blows. Sparks fly	>200	>12
Very Strong	Lumps only chip by heavy hammer blows. Dull ringing sound	100-200	6-12
Strong	Lumps or core broken by heavy hammer blow	50-100	3-6
Moderately Strong	Lump or core broken by light hammer blow	12.5-50	0.75-3
Moderately weak	Thin slabs broken by heavy hand pressure	5-12.5	0.3-0.75
Weak	Thin slabs break easily in hand	1.25-5	0.075-0.3
Very Weak	Crumbles in hand	<1.25	<0.075
Very Stiff	Can be indented by thumb nail	>0.3	

MATERIALS STORAGE AND SAFETY

MATERIALS MANAGEMENT ON CONSTRUCTION SITES

Materials management involves procedures and practices designed to reduce or eliminate pollution of storm water from stored materials.

Environmental stewardship entails use of management procedures for stockpiling:

- Contaminated and uncontaminated soil.
- Vegetative waste and paving materials.

- Materials removed from drains, ditches and culverts.
- Waste piles and any other material that could impact stormwater quality.

The following environmental stewardship practices are recommended in managing materials at construction sites:

- Minor slides/slip outs often occur during major storms. Stockpiles should be removed as soon as practicable and materials should be placed so that waterways are not impacted. See sediment control measures.
- During rain events, stockpiles of "cold mix" asphalt (*i.e.*, pre-mixed aggregate and asphalt binder) should be covered. During rain events, soil stockpiles should be covered or protected with soil stabilization measures and a temporary perimeter sediment barrier.
- During rain events, stockpiles of Portland cement concrete rubble, asphalt concrete, asphalt concrete rubble, aggregate base or aggregate subbase should be covered or protected with a temporary perimeter sediment barrier.

MATERIALS STORAGE

- Sites where chemicals, cements, solvents, paints, or other potential water pollutants are to be stored should be isolated in areas where they will not cause run-off pollution.
- Toxic chemicals and materials, such as pesticides, paints, and acids, should be stored according to manufacturers' guidelines.
- Overuse of chemicals should be avoided and great care should be taken to prevent accidental spillage.
- Containers should never be washed in or near flowing streams or storm water conveyance systems.
- Groundwater resources should be protected from leaching by placing a plastic mat, tar paper, or other impervious materials on any areas where toxic liquids are to be opened and stored.

Control of Sediment from Raw Materials Storage Areas

Caltrans has recommended the following stewardship practices for control of sediment from raw material storage areas:

- Water quality, erosion and sediment control BMPs should be properly implemented and regularly maintained.
- Wind erosion control practices should be implemented as appropriate on all stockpiled material.
- In general, stockpiles should be covered or protected with a temporary perimeter sediment barrier at all times. Perimeter controls and covers should be repaired and/or replaced as needed to keep them functioning properly.
- Berms should be installed around storage areas to minimize tracking of materials out of storage areas and to contain sediment within the storage area. Permanent rolled berms or ramp berms should be made of hot asphalt or Portland Concrete Cement (PCC). Cold mix asphalt is not recommended for use as raw material containment berms. Over time, cold mix has the potential to break up and not function as well as hot mix asphalt or PCC. Cover raw materials (especially cold mix) during the rainy season and have covers readily available outside the rainy season when rain is predicted.
- Sweep surfaces where material is tracked, blown, spilled or washed from the storage area.
- Reduce the size of stockpiles or the amount of stockpiled materials during the rainy season.
- Interim sediment controls include using temporary sediment controls to contain raw materials such as sand bags, straw bales or silt fences. Temporary sediment controls, such as sand bags and straw bales can degrade and may contribute to storm water pollution. Temporary and permanent sediment controls should be inspected regularly and replaced or repaired as needed.

- Both permanent and temporary sediment controls require maintenance. Sediment contained by temporary or permanent controls should be removed periodically.

POTENTIAL HAZARDS

A construction site has many materials at any one point. Most of these are usually in their raw state, meaning that they will undergo some process before they can be input into the building to form a part of the building.

They come in different forms and can be categorized as below:

- *Factory Goods*: These are mostly off the shelf items, they are unique in the fact that they can easily be resold and are therefore easy targets for pilferage. They are also delicate in the fact that they have unique storage requirements. Examples here include;
  - *Cement*: The most important attribute to consider in the storage of cement is the fact that it reacts chemically when in contact with moisture. For this reason, it should be kept under shade and on a platform, away from excessive moisture.
  - *Ceramics*: These include water closets, wash basins, Tiles and the like. They are extremely delicate and will easily break. This attribute is also shared with glasses. They should therefore be properly packaged in padded cartons and away from areas of much activity, usually under lock and key.
  - *Ironmongery*: These include locks, hinges, handles and the like. Owing to their small sizes, they are prone to pilfering. These should also be kept well locked and only issued under strict accountability.
- *Raw materials*: This category belongs to the main items like stone, ballast and sand. These are not prone to the previous problems like weather and pilferage. However, they have one attribute that is being bulky. They consume a lot of space on site and require a generous allocation of storage space. These are best

stored in bays and contained using things like stones, in the case of sand.

- *Workshop finished items*: This category also includes semi-finished items, for example in the case of timber. Items here are usually ready for installing in the works and are mostly purpose made. Some may have been imported from overseas and in their exact measurements. This means that damage or loss of such will lead to a very expensive work of replacement. Examples here include Fixtures, timber, roofing materials et cetera.

Material storage is a very important part of site management. How materials are delivered and dispatched determines how easily things flow. In almost all cases, site space is usually restricted and as such, material storage should be very well thought out. One thing to consider is that only the important and requisite materials and items per time should be stored on site to minimize on the risks mentioned above. Records should be kept very in good accuracy of all materials required, ordered, delivered, accepted, stored, dispatched, put to the works and any deficits.

Good material storage ensures the following benefits on site:

- Easy movement around site
- Reduction in waste and damages
- Reduction of loss by theft or otherwise

Handling and storing materials involves diverse operations such as hoisting tons of steel with a crane, driving a truck loaded with concrete blocks, manually carrying bags and material, and stacking drums, barrels, kegs, lumber, or loose bricks.

The efficient handling and storing of materials is vital to industry. These operations provide a continuous flow of raw materials, parts, and assemblies through the workplace, and ensure that materials are available when needed. Yet, the improper handling and storing of materials can cause costly injuries.

Workers frequently cite the weight and bulkiness of objects being lifted as major contributing factors to their injuries. In

1990, back injuries resulted in 400,000 workplace accidents. The second factor frequently cited by workers as contributing to their injuries was body movement. Bending, followed by twisting and turning, were the more commonly cited movements that caused back injuries. Back injuries accounted for more than 20 per cent of all occupational illnesses, according to data from the National Safety Council. In addition, workers can be injured by falling objects, improperly stacked materials, or by various types of equipment.

When manually moving materials, however, workers should be aware of potential injuries, including the following:

- Strains and sprains from improperly lifting loads, or from carrying loads that are either too large or too heavy.
- Fractures and bruises caused by being struck by materials, or by being caught in pinch points; and
- Cuts and bruises caused by falling materials that have been improperly stored, or by incorrectly cutting ties or other securing devices.

Since numerous injuries can result from improperly handling and storing materials, it is important to be aware of accidents that may occur from unsafe or improperly handled equipment and improper work practices, and to recognize the methods for eliminating, or at least minimizing, the occurrence of those accidents. Consequently, employers and employees can and should examine their workplaces to detect any unsafe or unhealthful conditions, practices, or equipment and take the necessary steps to correct them.

METHODS OF PREVENTION

General safety principles can help reduce workplace accidents. These include work practices, ergonomic principles, and training and education. Whether moving materials manually or mechanically, employees should be aware of the potential hazards associated with the task at hand and know how to exercise control over their workplaces to minimize the danger.

MOVING, HANDLING, AND STORING MATERIALS

When manually moving materials, employees should seek help when a load is so bulky it cannot be properly grasped or lifted, when they cannot see around or over it, or when a load cannot be safely handled.

When an employee is placing blocks under raised loads, the employee should ensure that the load is not released until his or her hands are clearly removed from the load. Blocking materials and timbers should be large and strong enough to support the load safely. Materials with evidence of cracks, rounded corners, splintered pieces, or dry rot should not be used for blocking.

Handles and holders should be attached to loads to reduce the chances of getting fingers pinched or smashed. Workers also should use appropriate protective equipment. For loads with sharp or rough edges, wear gloves or other hand and forearm protection. To avoid injuries to the hands and eyes, use gloves and eye protection. When the loads are heavy or bulky, the mover should also wear steel-toed safety shoes or boots to prevent foot injuries if the worker slips or accidentally drops a load.

When mechanically moving materials, avoid overloading the equipment by letting the weight, size, and shape of the material being moved dictate the type of equipment used for transporting it. All materials handling equipment has rated capacities that determine the maximum weight the equipment can safely handle and the conditions under which it can handle those weights. The equipment-rated capacities must be displayed on each piece of equipment and must not be exceeded except for load testing. When picking up items with a powered industrial truck, the load must be centered on the forks and as close to the mast as possible to minimize the potential for the truck tipping or the load falling. A lift truck must never be overloaded because it would be hard to control and could easily tip over. Extra weight must not be placed on the rear of a counterbalanced forklift to offset an overload. The load must be at the lowest position for traveling, and the truck manufacturer's operational requirements must be followed. All stacked loads must be correctly piled and cross-tiered, where possible. Precautions also should be taken when stacking and storing material.

Stored materials must not create a hazard. Storage areas must be kept free from accumulated materials that may cause tripping, fires, or explosions, or that may contribute to the harboring of rats and other pests. When stacking and piling materials, it is important to be aware of such factors as the materials' height and weight, how accessible the stored materials are to the user, and the condition of the containers where the materials are being stored.

All bound material should be stacked, placed on racks, blocked, interlocked, or otherwise secured to prevent it from sliding, falling, or collapsing. A load greater than that approved by a building official may not be placed on any floor of a building or other structure. Where applicable, load limits approved by the building inspector should be conspicuously posted in all storage areas.

When stacking materials, height limitations should be observed. For example, lumber must be stacked no more than 16 feet high if it is handled manually; 20 feet is the maximum stacking height if a forklift is used. For quick reference, walls or posts may be painted with stripes to indicate maximum stacking heights.

Used lumber must have all nails removed before stacking. Lumber must be stacked and leveled on solidly supported bracing. The stacks must be stable and self-supporting. Stacks of loose bricks should not be more than 7 feet in height. When these stacks reach a height of 4 feet, they should be tapered back 2 inches for every foot of height above the 4-foot level. When masonry blocks are stacked higher than 6 feet, the stacks should be tapered back one-half block for each tier above the 6-foot level.

Bags and bundles must be stacked in interlocking rows to remain secure. Bagged material must be stacked by stepping back the layers and cross-keying the bags at least every ten layers. To remove bags from the stack, start from the top row first. Baled paper and rags stored inside a building must not be closer than 18 inches to the walls, partitions, or sprinkler heads. Boxed materials must be banded or held in place using cross-ties or shrink plastic fibre.

Drums, barrels, and kegs must be stacked symmetrically. If stored on their sides, the bottom tiers must be blocked to keep them from rolling. When stacked on end, put planks, sheets of plywood dunnage, or pallets between each tier to make a firm, flat, stacking surface. When stacking materials two or more tiers high, the bottom tier must be chocked on each side to prevent shifting in either direction.

When stacking, consider the need for availability of the material. Material that cannot be stacked due to size, shape, or fragility can be safety stored on shelves or in bins. Structural steel, bar stock, poles, and other cylindrical materials, unless in racks, must be stacked and blocked to prevent spreading or tilting. Pipes and bars should not be stored in racks that face main aisles; this could create a hazard to passers-by when supplies are being removed.

USING MATERIALS HANDLING EQUIPMENT

To reduce potential accidents associated with workplace equipment, employees need to be trained in the proper use and limitations of the equipment they operate. This includes knowing how to effectively use equipment such as conveyors, cranes, and slings.

Conveyors

When using conveyors, workers' hands may be caught in nip points where the conveyor runs over support members or rollers; workers may be struck by material falling off the conveyor; or they may become caught on or in the conveyor, thereby being drawn into the conveyor path.

To reduce the severity of an injury, an emergency button or pull cord designed to stop the conveyor must be installed at the employee's work station. Continuously accessible conveyor belts should have an emergency stop cable that extends the entire length of the conveyor belt so that the cable can be accessed from any location along the belt. The emergency stop switch must be designed to be reset before the conveyor can be restarted. Before restarting a conveyor that has stopped due to an overload, appropriate personnel must inspect the conveyor and clear the stoppage before restarting. Employees must never ride on a materials handling conveyor. Where a conveyor passes over work areas or aisles, guards must be provided to keep employees from being struck by falling material. If the crossover is low enough for workers to run into, it must be guarded to protect employees and either marked with a warning sign or painted a bright colour.

Screw conveyors must be completely covered except at loading and discharging points. At those points, guards must protect employees against contacting the moving screw; the guards are movable, and they must be interlocked to prevent conveyor movement when not in place.

Cranes

Only thoroughly trained and competent persons are permitted to operate cranes. Operators should know what they are lifting and what it weighs. The rated capacity of mobile cranes varies with the length of the boom and the boom radius. When a crane has a telescoping boom, a load may be safe to lift at a short boom length and/or a short boom radius, but may overload the crane when the boom is extended and the radius increases.

All movable cranes must be equipped with a boom angle indicator; those cranes with telescoping booms must be equipped with some means to determine the boom length, unless the load rating is independent of the boom length. Load rating charts must be posted in the cab of cab-operated cranes. All mobile cranes do not have uniform capacities for the same boom length and radius in all directions around the chassis of the vehicle. Always check the crane's load chart to ensure that the crane is not going to be overloaded for the conditions under which it will operate. Plan lifts before starting them to ensure that they are safe. Take additional precautions and exercise extra care when operating around power lines.

Some mobile cranes cannot operate with outriggers in the traveling position. When used, the outriggers must rest on firm ground, on timbers, or be sufficiently cribbed to spread the weight of the crane and the load over a large enough area. This will prevent the crane from tipping during use. Hoisting chains and ropes must always be free of kinks or twists and must never be wrapped around a load. Loads should be attached to the load hook by slings, fixtures, or other devices that have the capacity to support the load on the hook. Sharp edges of loads should be padded to prevent cutting slings. Proper sling angles shall be maintained so that slings are not loaded in excess of their capacity.

All cranes must be inspected frequently by persons thoroughly familiar with the crane, the methods of inspecting the crane, and what can make the crane unserviceable. Crane activity, the severity of use, and environmental conditions should determine inspection schedules. Critical parts, such as crane operating mechanisms, hooks, air or hydraulic system components and other load-carrying components, should be inspected daily for any maladjustment, deterioration, leakage, deformation, or other damage.

Slings

When working with slings, employers must ensure that they are visually inspected before use and during operation, especially if used under heavy stress. Riggers or other knowledgeable employees should conduct or assist in the inspection because they are aware of how the sling is used and what makes a sling unserviceable. A damaged or defective sling must be removed from service.

Slings must not be shortened with knots or bolts or other makeshift devices, sling legs that have been kinked must not be used. Slings must not be loaded beyond their rated capacity, according to the manufacturer's instructions. Suspended loads must be kept clear of all obstructions, and crane operators should avoid sudden starts and stops when moving suspended loads. Employees also must remain clear of loads about to be lifted and suspended. All shock loading is prohibited.

Workers who must handle and store materials often use fork trucks, platform lift trucks, motorized hand trucks, and other specialized industrial trucks powered by electrical motors or internal combustion engines. Affected workers, therefore, should be aware of the safety requirements pertaining to fire protection, and the design, maintenance, and use of these trucks.

All new powered industrial trucks, except vehicles intended primarily for earth moving or over-the-road hauling, shall meet the design and construction requirements for powered industrial trucks established in the American National Standard for Powered Industrial Trucks, Part ll, ANSI B56.1-1969. Approved trucks shall also bear a label or some other identifying mark indicating acceptance by a nationally recognized testing laboratory.

Modifications and additions that affect capacity and safe operation of the trucks shall not be performed by an owner or user without the manufacturer's prior written approval. In these cases, capacity, operation, and maintenance instruction plates and tags or decals must be changed to reflect the new information.

If the truck is equipped with front-end attachments that are not factory installed, the user should request that the truck be marked to identify these attachments and show the truck's approximate weight, including the installed attachment, when it is at maximum elevation with its load laterally centered.

There are 11 different types of industrial trucks or tractors, some having greater safeguards than others. There are also designated conditions and locations under which the vast range of industrial-powered trucks can be used. In some instances, powered industrial trucks cannot be used, and in others, they can only be used if approved by a nationally recognized testing laboratory for fire safety.

For example, powered industrial trucks must not be used in atmospheres containing hazardous concentrations of the following substances:

- Acetylene
- Butadiene
- Ethylene oxide
- Hydrogen (or gases or vapors equivalent in hazard to hydrogen, such as manufactured gas)
- Propylene oxide
- Acetaldehyde
- Cyclopropane
- Dimethyl ether
- Ethylene
- Isoprene, and
- Unsymmetrical dimethyl hydrazine

These trucks are not to be used in atmospheres containing hazardous concentrations of metal dust, including aluminum, magnesium, and other metals of similarly hazardous characteristics or in atmospheres containing carbon black, coal, or coke dust. Where dust of magnesium, aluminum, or aluminum bronze dusts may be present, the fuses, switches, motor controllers, and circuit breakers of trucks must be enclosed with enclosures approved for these substances.

There also are powered industrial trucks or tractors that are designed, constructed, and assembled for use in atmospheres containing flammable vapors or dusts. These include industrial-powered trucks equipped with additional safeguards to their exhaust, fuel, and electrical systems; with no electrical equipment, including the ignition; with temperature limitation features; and with electric motors and all other electrical equipment completely enclosed.

These specially designed powered industrial trucks may be used in locations where volatile flammable liquids or flammable gases are handled, processed, or used. The liquids, vapors, or gases should, among other things, be confined within closed containers or. closed systems from which they cannot escape.

Some other conditions and/or locations in which specifically designed powered industrial trucks may be used include the following:

- Only powered industrial trucks that do not have any electrical equipment, including the ignition, and have their electrical motors or other electrical equipment completely enclosed should be used in atmospheres containing flammable vapors or dust.
- Powered industrial trucks that are either powered electrically by liquified petroleum gas or by a gasoline or diesel engine are used on piers and wharves that handle general cargo.

Safety precautions the user can observe when operating or maintaining powered industrial trucks include:

- That high lift rider trucks be fitted with an overhead guard, unless operating conditions do not permit.
- That fork trucks be equipped with a vertical load backrest extension according to manufacturers' specifications, if the load presents a hazard.
- That battery charging installations be located in areas designated for that purpose.
- That facilities be provided for flushing and neutralizing spilled electrolytes when changing or recharging a battery to prevent fires, to protect the charging apparatus from being damaged by the trucks, and to adequately ventilate fumes in the charging area from gassing batteries.
- That conveyor, overhead hoist, or equivalent materials handling equipment be provided for handling batteries.
- That auxiliary directional lighting be provided on the truck where general lighting is less than 2 lumens per square foot.
- That arms and legs not be placed between the uprights of the mast or outside the running lines of the truck.
- That brakes be set and wheel blocks or other adequate protection be in place to prevent movement of trucks, trailers, or railroad cars when using trucks to load or unload materials onto train boxcars.

- That sufficient headroom be provided under overhead installations, lights, pipes, and sprinkler systems.
- That personnel on the loading platform have the means to shut off power to the truck.
- That dockboards or bridgeplates be properly secured, so they won't move when equipment moves over them.
- That only stable or safely arranged loads be handled, and caution be exercised when handling loads.
- That trucks whose electrical systems are in need of repair have the battery disconnected prior to such repairs.
- That replacement parts of any industrial truck be equivalent in safety to the original ones.

ERGONOMIC SAFETY AND HEALTH PRINCIPLES

Ergonomics is defined as the study of work and is based on the principle that the job should be adapted to fit the person, rather than forcing the person to fit the job. Ergonomics focuses on the work environment and items such as design and function of workstations, controls, displays, safety devices, tools, and lighting to fit the employees' physical requirements and to ensure their health and well being.

Ergonomics includes restructuring or changing workplace conditions to make the job easier and reducing/stressors that cause cumulative trauma disorders and repetitive motion injuries. In the area of materials handling and storing, ergonomic principles may require controls such as reducing the size or weight of the objects lifted, installing a mechanical lifting aid, or changing the height of a pallet or shelf.

Although no approach has been found for totally eliminating back injuries resulting from lifting materials, a substantial number of lifting injuries can be prevented by implementing an effective ergonomics programme and by training employees in appropriate lifting techniques.

In addition to using ergonomic controls, there are some basic safety principles that can be employed to reduce injuries resulting from handling and storing materials. These include taking general fire safety precautions and keeping aisles and passageways clear. In adhering to fire safety precautions, employees should note that flammable and combustible materials must be stored according to their fire characteristics. Flammable liquids, for example, must be separated from other material by a fire wall. Also, other combustibles must be stored in an area where smoking and using an open flame or a spark-producing device is prohibited. Dissimilar materials that are dangerous when they come into contact with each other must be stored apart.

When using aisles and passageways to move materials mechanically, sufficient clearance must be allowed for aisles at loading docks, through doorways, wherever turns must be made, and in other parts of the workplace. Providing sufficient clearance for mechanically moved materials will prevent workers from being pinned between the equipment and fixtures in the workplace, such as walls, racks, posts, or other machines. Sufficient clearance also will prevent the load from striking an obstruction and falling on an employee. All passageways used by employees should be kept clear of obstructions and tripping hazards. Materials in excess of supplies needed for immediate operations should not be stored in aisles or passageways, and permanent aisles and passageways must be marked appropriately.

TRAINING AND EDUCATION

OSHA recommends using a formal training programme to reduce materials handling hazards. Instructors should be well-versed in matters that pertain to safety engineering and materials handling and storing.

The content of the training should emphasize those factors that will contribute to reducing workplace hazards including the following:

- Alerting the employee to the dangers of lifting without proper training.
- Showing the employee how to avoid unnecessary physical stress and strain.
- Teaching workers to become aware of what they can comfortably handle without undue strain.

- Instructing workers on the proper use of equipment.
- Teaching workers to recognize potential hazards and how to prevent or correct them.

Because of the high incidence of back injuries, safe lifting techniques for manual lifting should be demonstrated and practiced at the work site by supervisors as well as by employees.

A training programme to teach proper lifting techniques should cover the following topics:

- Awareness of the health risks to improper lifting—citing organizational case histories.
- Knowledge of the basic anatomy of the spine, the muscles, and the joints of the trunk, and the contributions of intra-abdominal pressure while lifting.
- Awareness of individual body strengths and weaknesses—determining one's own lifting capacity.
- Recognition of the physical factors that might contribute to an accident, and how to avoid the unexpected.
- Use of safe lifting postures and timing for smooth, easy lifting and the ability to minimize the load-moment effects.
- Use of handling aids such as stages, platforms, or steps, trestles, shoulder pads, handles, and wheels.
- Knowledge of body responses—warning signals—to be aware of when lifting.

A campaign using posters to draw attention to the need to do something about potential accidents, including lifting and back injuries, is one way to increase awareness of safe work practices and techniques. The plant medical staff and a team of instructors should conduct regular tours of the site to look for potential hazards and allow input from workers.

SAFETY AND HEALTH PROGRAMME MANAGEMENT GUIDELINES

To have an effective materials handling and storing safety and health programme, managers must take an active role in its development. First-line supervisors must be convinced of the importance of controlling hazards associated with materials handling and storing and must be held accountable for employee training. An on-going safety and health programme should be used to motivate employees to continue to use necessary protective gear and to observe proper job procedures.

OSHA's recommended "Safety and Health Programme Management Guidelines" issued in 1989 can provide a blueprint for employers who are seeking guidance on how to effectively manage and protect worker safety and health.

The four main elements of an effective occupational safety and health programme are:

1. Management commitment and employee involvement,
2. Worksite analysis,
3. Hazard prevention and control, and
4. Safety and health training.

These elements encompass principles such as establishing and communicating clear goals of a safety and health management programme; conducting worksite examinations to identify existing hazards and the conditions under which changes might occur; effectively designing the job site or job to prevent hazards; and providing essential training to address the safety and health responsibilities of both management and employees.

6

Labour, Material and Equipment

HISTORICAL PERSPECTIVE

Good project management in construction must vigorously pursue the efficient utilization of labor, material and equipment. Improvement of labor productivity should be a major and continual concern of those who are responsible for cost control of constructed facilities. Material handling, which includes procurement, inventory, shop fabrication and field servicing, requires special attention for cost reduction. The use of new equipment and innovative methods has made possible wholesale changes in construction technologies in recent decades. Organizations which do not recognize the impact of various innovations and have not adapted to changing environments have justifiably been forced out of the mainstream of construction activities.

Observing the trends in construction technology presents a very mixed and ambiguous picture. On the one hand, many of the techniques and materials used for construction are essentially unchanged since the introduction of mechanization in the early part of the twentieth century. For example, a history of the Panama Canal construction from 1904 to 1914 argues that: The work could not have done any faster or more efficiently in our day, despite all technological and mechanical advances in the time since, the reason being that no present system could possibly carry the spoil away any faster or more efficiently than the system employed. No motor trucks were used in the digging of the canal; everything ran on rails. And because of the mud and rain, no other method would have worked half so well.

In contrast to this view of one large project, one may also point to the continual change and improvements occurring in traditional materials and techniques. Bricklaying provides a good example of such changes:

Bricklaying...is said not to have changed in thousands of years; perhaps in the literal placing of brick on brick it has not. But masonry technology has changed a great deal. Motorized wheelbarrows and mortar mixers, sophisticated scaffolding systems, and forklift trucks now assist the bricklayer. New epoxy mortars give stronger adhesion between bricks. Mortar additives and cold-weather protection eliminate winter shutdowns.

Add to this list of existing innovations the possibility of robotic bricklaying; automated prototypes for masonry construction already exist. Technical change is certainly occurring in construction, although it may occur at a slower rate than in other sectors of the economy.

The United States construction industry often points to factors which cannot be controlled by the industry as a major explanatory factor in cost increases and lack of technical innovation. These include the imposition of restrictions for protection of the environment and historical districts, requirements for community participation in major construction projects, labor laws which allow union strikes to become a source of disruption, regulatory policies including building codes and zoning ordinances, and tax laws which inhibit construction abroad. However, the construction industry should bear a large share of blame for not realizing earlier that the technological edge held by the large U.S. construction firms has eroded in face of stiff foreign competition. Many past practices, which were tolerated when U.S. contractors had a technological lead, must now be changed in the face of stiff competition. Otherwise, the U.S. construction industry will continue to find itself in trouble.

With a strong technological base, there is no reason why the construction industry cannot catch up and reassert itself to meet competition wherever it may be. Individual design and/or construction firms must explore new ways to improve productivity for the future. Of course, operational planning for construction projects is still important, but such tactical planning has limitations and may soon reach the point of diminishing return because much that can be wrung out of the existing practices have already been tried. What is needed the most is strategic planning to usher in a revolution which can improve productivity by an order of magnitude or more. Strategic planning should look at opportunities and ask whether there are potential options along which new goals may be sought on the basis of existing resources. No one can be certain about the success of various development options for the design professions and the construction industry. However, with the availability of today's high technology, some options have good potential of success because of the social and economic necessity which will eventually push barriers aside. Ultimately, decisions for action, not plans, will dictate future outcomes.

LABOR PRODUCTIVITY

Productivity in construction is often broadly defined as output per labor hour. Since labor constitutes a large part of the construction cost and the quantity of labor hours in performing a task in construction is more susceptible to the influence of management than are materials or capital, this productivity measure is often referred to as *labor productivity*. However, it is important to note that labor productivity is a measure of the overall effectiveness of an operating system in utilizing labor, equipment and capital to convert labor efforts into useful output, and is not a measure of the capabilities of labor alone. For example, by investing in a piece of new equipment to perform certain tasks in construction, output may be increased for the same number of labor hours, thus resulting in higher labor productivity.

Construction output may be expressed in terms of functional units or constant dollars. In the former case, labor productivity is associated with units of product per labor hour, such as cubic yards of concrete placed per hour or miles of highway paved per hour. In the latter case, labor productivity is identified with value of construction (in constant dollars) per labor hour. The value of construction in this regard is not measured by the benefit of constructed facilities, but by construction cost. Labor productivity measured in this way requires considerable care in interpretation. For example, wage rates in construction have been declining in the US during the period 1970 to 1990, and since wages are an important component in construction costs, the value of construction put in place per hour of work will decline as a result, suggesting lower productivity.

PRODUCTIVITY AT THE JOB SITE

Contractors and owners are often concerned with the labor activity at job sites. For this purpose, it is convenient to express labor productivity as functional units per labor hour for each type of construction task. However, even for such specific purposes, different levels of measure may be used. For example, cubic yards of concrete placed per hour is a lower level of measure than miles of highway paved per hour. Lower-level measures are more useful for monitoring individual activities, while higher-level measures may be more convenient for developing industry-wide standards of performance.

While each contractor or owner is free to use its own system to measure labor productivity at a site, it is a good practice to set up a system which can be used to track productivity trends over time and in varied locations. Considerable efforts are required to collect information regionally or nationally over a number of years to produce such results. The productivity indices compiled from statistical data should include parameters such as the performance of major crafts, effects of project size, type and location, and other major project influences.

In order to develop industry-wide standards of performance, there must be a general agreement on the measures to be useful for compiling data. Then, the job site productivity data collected by various contractors and owners can be correlated and analysed to develop certain measures for each of the major segment of the construction industry. Thus, a contractor or owner can compare its performance with that of the industry average.

PRODUCTIVITY IN THE CONSTRUCTION INDUSTRY

Because of the diversity of the construction industry, a single index for the entire industry is neither meaningful nor reliable. Productivity indices may be developed for major segments of the construction industry nationwide if reliable statistical data can be obtained for separate industrial segments. For this general type of productivity measure, it is more convenient to express labor productivity as constant dollars per labor hours since dollar values are more easily aggregated from a large amount of data collected from different sources. The use of constant dollars allows meaningful approximations of the changes in construction output from one year to another when price deflators are applied to current dollars to obtain the corresponding values in constant dollars. However, since most construction price deflators are obtained from a combination of price indices for material and labor inputs, they reflect only the change of price levels and do not capture any savings arising from improved labor productivity. Such deflators tend to overstate increases in construction costs over a long period of time, and consequently understate the physical volume or value of construction work in years subsequent to the base year for the indices.

FACTORS AFFECTING JOB-SITE PRODUCTIVITY

Job-site productivity is influenced by many factors which

can be characterized either as labor characteristics, project work conditions or as non-productive activities.

The labor characteristics include:

- Age, skill and experience of workforce
- Leadership and motivation of workforce

The project work conditions include among other factors:

- Job size and complexity.
- Job site accessibility.
- Labor availability.
- Equipment utilization.
- Contractual agreements.
- Local climate.
- Local cultural characteristics, particularly in foreign operations.

The non-productive activities associated with a project may or may not be paid by the owner, but they nevertheless take up potential labor resources which can otherwise be directed to the project.

The non-productive activities include among other factors:

- Indirect labor required to maintain the progress of the project
- Rework for correcting unsatisfactory work
- Temporary work stoppage due to inclement weather or material shortage
- Time off for union activities
- Absentee time, including late start and early quits
- Non-working holidays
- Strikes

Each category of factors affects the productive labor available to a project as well as the on-site labor efficiency.

LABOR CHARACTERISTICS

Performance analysis is a common tool for assessing worker quality and contribution.

Factors that might be evaluated include:

- *Quality of Work*: Caliber of work produced or accomplished.

- *Quantity of Work*: Volume of acceptable work
- *Job Knowledge*: Demonstrated knowledge of requirements, methods, techniques and skills involved in doing the job and in applying these to increase productivity.
- *Related Work Knowledge*: Knowledge of effects of work upon other areas and knowledge of related areas which have influence on assigned work.
- *Judgment*: Soundness of conclusions, decisions and actions.
- *Initiative*: Ability to take effective action without being told.
- *Resource Utilization*: Ability to delineate project needs and locate, plan and effectively use all resources available.
- *Dependability*: Reliability in assuming and carrying out commitments and obligations.
- *Analytical Ability*: Effectiveness in thinking through a problem and reaching sound conclusions.
- *Communicative Ability*: Effectiveness in using orgal and written communications and in keeping subordinates, associates, superiors and others adequately informed.
- *Interpersonal Skills*: Effectiveness in relating in an appropriate and productive manner to others.
- *Ability to Work Under Pressure*: Ability to meet tight deadlines and adapt to changes.
- *Security Sensitivity*: Ability to handle confidential information appropriately and to exercise care in safeguarding sensitive information.
- *Safety Consciousness*: Has knowledge of good safety practices and demonstrates awareness of own personal safety and the safety of others.
- *Profit and Cost Sensitivity*: Ability to seek out, generate and implement profit-making ideas.
- *Planning Effectiveness*: Ability to anticipate needs, forecast conditions, set goals and standards, plan and schedule work and measure results.

- *Leadership*: Ability to develop in others the willingenss and desire to work towards common objectives.
- *Delegating*: Effectiveness in delegating work appropriately.
- *Development People*: Ability to select, train and appraise personnel, set standards of performance, and provide motivation to grow in their capacity. < li>Diversity (Equal Employment Opportunity)-ability to be senstive to the needs of minorities, females and other protected groups and to demonstrate affirmative action in responding to these needs.

These different factors could each be assessed on a three point scale:

- Recognized strength,
- Meets expectations,
- Area needing improvement.

Examples of work performance in these areas might also be provided.

PROJECT WORK CONDITIONS

Job-site labor productivity can be estimated either for each craft (carpenter, bricklayer, etc.) or each type of construction (residential housing, processing plant, etc.) under a specific set of work conditions. A *base labor productivity* may be defined for a set of work conditions specified by the owner or contractor who wishes to observe and measure the labor performance over a period of time under such conditions. A *labor productivity index* may then be defined as the ratio of the job-site labor productivity under a different set of work conditions to the base labor productivity, and is a measure of the relative labor efficiency of a project under this new set of work conditions.

The effects of various factors related to work conditions on a new project can be estimated in advance, some more accurately than others. For example, for very large construction projects, the labor productivity index tends to decrease as the project size and/or complexity increase because of logistic problems and the "learning" that the work force must undergo before adjusting to the new environment. Job-site accessibility often may reduce the labor productivity index if the workers must perform their jobs in round about ways, such as avoiding traffic in repaving the highway surface or maintaining the operation of a plant during renovation. Labor availability in the local market is another factor. Shortage of local labor will force the contractor to bring in non-local labor or schedule overtime work or both. In either case, the labor efficiency will be reduced in addition to incurring additional expenses.

The degree of equipment utilization and mechanization of a construction project clearly will have direct bearing on job-site labor productivity. The contractual agreements play an important role in the utilization of union or non-union labor, the use of subcontractors and the degree of field supervision, all of which will impact job-site labor productivity. Since on-site construction essentially involves outdoor activities, the local climate will influence the efficiency of workers directly. In foreign operations, the cultural characteristics of the host country should be observed in assessing the labor efficiency.

NON-PRODUCTIVE ACTIVITIES

The non-productive activities associated with a project should also be examined in order to examine the *productive labor yield,* which is defined as the ratio of direct labor hours devoted to the completion of a project to the potential labor hours. The direct labor hours are estimated on the basis of the best possible conditions at a job site by excluding all factors which may reduce the productive labor yield. For example, in the repaving of highway surface, the flagmen required to divert traffic represent indirect labor which does not contribute to the labor efficiency of the paving crew if the highway is closed to the traffic. Similarly, for large projects in remote areas, indirect labor may be used to provide housing and infrastructure for the workers hired to supply the direct labor for a project. The labor hours spent on rework to correct unsatisfactory original work represent extra time taken away from potential labor hours. The labor hours related to such activities must be deducted from the potential labor hours in order to obtain the actual productive labor yield.

Effects of Job Size on Productivity

A contractor has established that under a set of "standard" work conditions for building construction, a job requiring 500,000 labor hours is considered standard in determining the base labor productivity. All other factors being the same, the labor productivity index will increase to 1.1 or 110% for a job requiring only 400,000 labor-hours. Assuming that a linear relation exists for the range between jobs requiring 300,000 to 700,000 labor hours as shown in Figure 6.1, determine the labor productivity index for a new job requiring 650,000 labor hours under otherwise the same set of work conditions.

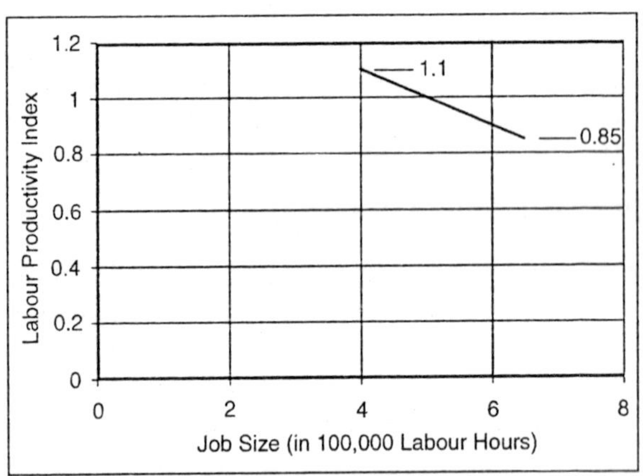

Fig. 6.1 Illustrative Relationship between Productivity Index and Job Size

The labor productivity index I for the new job can be obtained by linear interpolation of the available data as follows:

$$I = 1.0 + (1.1 - 1.0)\left[\frac{500{,}000 - 650{,}000}{500{,}000 - 400{,}000}\right] = 0.85$$

This implies that labor is 15% less productive on the large job than on the standard project.

Productive Labor Yield

In the construction of an off-shore oil drilling platform, the potential labor hours were found to be L = 7.5 million hours.

Of this total, the non-productive activities expressed in thousand labor hours were as follows:

- A = 417 for holidays and strikes
- B = 1,415 for absentees (*i.e.* vacation, sick time, etc.)
- C = 1,141 for temporary stoppage (*i.e.* weather, waiting, union activities, etc.)
- D = 1,431 for indirect labor (*i.e.* building temporary facilities, cleaning up the site, rework to correct errors, etc.)

Determine the productive labor yield after the above factors are taken into consideration.

The percentages of time allocated to various non-productive activities, A, B, C and D are:

$$\frac{A}{L} = \frac{417}{7,500} = 6\% \qquad \frac{B}{L} = \frac{1,415}{7,500} = 19\%$$

$$\frac{C}{L} = \frac{1,141}{7,500} = 15\% \qquad \frac{D}{L} = \frac{1,431}{7,500} = 19\%$$

The total percentage of time X for all non-productive activities is:

$$X = \frac{A + B + C + D}{L} = 6\% + 19\% + 15\% + 19\% = 59\%$$

The productive labor yield, Y, when the given factors for A, B, C and D are considered, is as follows:

$$Y = \frac{L - A - B - C - D}{L} = 100\% - 6\% - 19\% - 15\% - 19\% = 41\%$$

As a result, only 41% of the budgeted labor time was devoted directly to work on the facility.

Utilization of On-site Worker's Time

An example illustrating the effects of indirect labor

requirements which limit productive labor by a typical craftsman on the job site was given by R. Tucker with the following percentages of time allocation:

Unproductive time	40%
Administrative delays	20%
Inefficient work methods	20%
Labor jurisdictions and other work restrictions	15%
Personal time	5%

In this estimate, as much time is spent on productive work as on delays due to management and inefficiencies due to antiquated work methods.

LABOR RELATIONS IN CONSTRUCTION

The market demand in construction fluctuates greatly, often within short periods and with uneven distributions among geographical regions. Even when the volume of construction is relatively steady, some types of work may decline in importance while other types gain. Under an unstable economic environment, employers in the construction industry place great value on flexibility in hiring and laying off workers as their volumes of work wax and wane. On the other hand, construction workers sense their insecurity under such circumstances and attempt to limit the impacts of changing economic conditions through labor organizations.

There are many crafts in the construction labor forces, but most contractors hire from only a few of these crafts to satisfy their specialized needs. Because of the peculiar characteristics of employment conditions, employers and workers are placed in a more intimate relationship than in many other industries. Labor and management arrangements in the construction industry include both unionized and non-unionized operations which compete for future dominance. Dramatic shifts in unionization can occur. For example, the fraction of trade union members in the construction industry declined from 42% in 1992 to 26% in 2000 in Australia, a 40% decline in 8 years.

UNIONIZED CONSTRUCTION

The craft unions work with construction contractors using unionized labor through various market institutions such as jurisdiction rules, apprenticeship programmes, and the referral system. Craft unions with specific jurisdiction rules for different trades set uniform hourly wage rates for journeymen and offer formal apprenticeship training to provide common and equivalent skill for each trade. Contractors, through the contractors' associations, enter into legally binding collective bargaining agreements with one or more of the craft unions in the construction trades. The system which binds both parties to a collective bargaining agreement is referred to as the "union shop". These agreements obligate a contractor to observe the work jurisdictions of various unions and to hire employees through a union operated referral system commonly known as the hiring hall.

The referral systems operated by union organizations are required to observe several conditions:

- All qualified workers reported to the referral system must be made available to the contractor without discrimination on the basis of union membership or other relationship to the union. The "closed shop" which limits referral to union members only is now illegal.
- The contractor reserves the right to hire or refuse to hire any worker referred by the union on the basis of his or her qualifications.
- The referral plan must be posted in public, including any priorities of referrals or required qualifications.

While these principles must prevail, referral systems operated by labor organizations differ widely in the construction industry. Contractors and craft unions must negotiate not only wage rates and working conditions, but also hiring and apprentice training practices. The purpose of trade jurisdiction is to encourage considerable investment in apprentice training on the part of the union so that the contractor will be protected by having only qualified workers perform the job even though such workers are not permanently attached to the contractor and thus may have no sense of security or loyalty. The referral system is often a rapid and dependable source of workers, particularly for a contractor who moves into a new geographical location or starts a new project which has high fluctuations in demand for labor. By and large, the referral system has functioned smoothly in providing qualified workers to contractors, even though some other aspects of union operations are not as well accepted by contractors.

NON-UNIONIZED CONSTRUCTION

In recent years, non-union contractors have entered and prospered in an industry which has a long tradition of unionization. Non-union operations in construction are referred to as "open shops." However, in the absence of collective bargaining agreements, many contractors operate under policies adopted by non-union contractors' associations. This practice is referred to as "merit shop", which follows substantially the same policies and procedures as collective bargaining although under the control of a non-union contractors' association without union participation. Other contractors may choose to be totally "unorganized" by not following either union shop or merit shop practices.

The operations of the merit shop are national in scope, except for the local or state apprenticeship and training plans. The comprehensive plans of the contractors' association apply to all employees and crafts of a contractor regardless of their trades. Under such operations, workers have full rights to move through the nation among member contractors of the association. Thus, the non-union segment of the industry is organized by contractors' associations into an integral part of the construction industry. However, since merit shop workers are employed directly by the construction firms, they have a greater loyalty to the firm, and recognize that their own interest will be affected by the financial health of the firm. Playing a significant role in the early growth and continued expansion of merit shop construction is the Associated Builders and Contractors association. By 1987, it had a membership of nearly 20,000 contractors and a network of 75 chapters through the nation. Among the merit shop contractors are large construction firms such as Fluor Daniel, Blount International, and Brown and Root Construction.

The advantages of merit shops as claimed by its advocates are:

- The ability to manage their own work force
- Flexibility in making timely management decisions
- The emphasis on making maximum usage of local labor force
- The emphasis on encouraging individual work advancement through continued development of skills
- The shared interest that management and workers have in seeing an individual firm prosper.

By shouldering the training responsibility for producing skill workers, the merit shop contractors have deflected the most serious complaints of users and labor that used to be raised against the open shop. On the other hand, the use of mixed crews of skilled workers at a job site by merit shop contractors enables them to remove a major source of inefficiencies caused by the exclusive jurisdiction practiced in the union shop, namely the idea that only members of a particular union should be permitted to perform any given task in construction. As a result, merit shop contractors are able to exert a beneficial influence on productivity and cost-effectiveness of construction projects.

The unorganized form of open shop is found primarily in housing construction where a large percentage of workers are characterized as unskilled helpers. The skilled workers in various crafts are developed gradually through informal apprenticeships while serving as helpers. This form of open shop is not expected to expand beyond the type of construction projects in which highly specialized skills are not required.

PROBLEMS IN COLLECTIVE BARGAINING

In the organized building trades in North American construction, the primary unit is the international union, which is an association of local unions in the United States and Canada. Although only the international unions have the power to issue or remove charters and to organize or combine local unions, each local union has considerable degrees of autonomy in the conduct of its affairs, including the negotiation of collective bargaining agreements. The business agent of a local union is an elected official who is the most important person in handling the day to day operations on behalf of the union. The contractors' associations representing the employers vary widely in composition and structure, particularly in different geographical regions. In general, local contractors' associations are considerably less well organized than the union with which they deal, but they try to strengthen themselves through affiliation with state and national organizations. Typically, collective bargaining agreements in construction are negotiated between a local union in a single craft and the employers of that craft as represented by a contractors' association, but there are many exceptions to this pattern. For example, a contractor may remain outside the association and negotiate independently of the union, but it usually cannot obtain a better agreement than the association.

Because of the great variety of bargaining structures in which the union and contractors' organization may choose to stage negotiations, there are many problems arising from jurisdictional disputes and other causes. Given the traditional rivalries among various crafts and the ineffective organization of some of contractors' associations, coupled with the lack of adequate mechanisms for settling disputes, some possible solutions to these problems deserve serious attention:

REGIONAL BARGAINING

Currently, the geographical area in a collective bargaining agreement does not necessarily coincide with the territory of the union and contractors' associations in the negotiations. There are overlapping of jurisdictions as well as territories, which may create successions of contract termination dates for different crafts. Most collective bargaining agreements are negotiated locally, but regional agreements with more comprehensive coverage embracing a number of states have been established. The role of national union negotiators and contractors' representatives in local collective bargaining is limited. The national agreement between international unions and a national contractor normally binds the contractors' association and its bargaining unit. Consequently, the most promising reform lies in the broadening of the geographic region of an agreement in a single trade without overlapping territories or jurisdictions.

MULTICRAFT BARGAINING

The treatment of interrelationships among various craft trades in construction presents one of the most complex issues in the collective bargaining process. Past experience on project agreements has dealt with such issues successfully in that collective bargaining agreements are signed by a group of craft trade unions and a contractor for the duration of a project. Project agreements may reference other agreements on particular points, such as wage rates and fringe benefits, but may set their own working conditions and procedures for settling disputes including a commitment of no-strike and no-lockout. This type of agreement may serve as a starting point for multicraft bargaining on a regional, non-project basis.

IMPROVEMENT OF BARGAINING PERFORMANCE

Although both sides of the bargaining table are to some degree responsible for the success or failure of negotiation, contractors have often been responsible for the poor performance of collective bargaining in construction in recent years because local contractors' associations are generally less well organized and less professionally staffed than the unions with which they deal. Legislation providing for contractors' association accreditation as an exclusive bargaining agent has now been provided in several provinces in Canada. It provides a government board that could hold hearings and establish an appropriate bargaining unit by geographic region or sector of the industry, on a single-trade or multi-trade basis.

MATERIALS MANAGEMENT

Materials management is an important element in project planning and control. Materials represent a major expense in construction, so minimizing *procurement* or*purchase* costs presents important opportunities for reducing costs. Poor materials management can also result in large and avoidable costs during construction. First, if materials are purchased early, capital may be tied up and interest charges incurred on the excess *inventory* of materials.

Even worse, materials may deteriorate during storage or be stolen unless special care is taken. For example, electrical equipment often must be stored in waterproof locations. Second, delays and extra expenses may be incurred if materials required for particular activities are not available. Accordingly, insuring a timely flow of material is an important concern of project managers. Materials management is not just a concern during the monitoring stage in which construction is taking place. Decisions about material procurement may also be required during the initial planning and scheduling stages. For example, activities can be inserted in the project schedule to represent purchasing of major items such as elevators for buildings. The availability of materials may greatly influence the schedule in projects with a *fast track* or very tight time schedule: sufficient time for obtaining the necessary materials must be allowed. In some case, more expensive suppliers or shippers may be employed to save time.

Materials management is also a problem at the organization level if central purchasing and inventory control is used for standard items. In this case, the various projects undertaken by the organization would present requests to the central purchasing group. In turn, this group would maintain inventories of standard items to reduce the delay in providing material or to obtain lower costs due to bulk purchasing. This organizational materials management problem is analogous to inventory control in any organization facing continuing demand for particular items. Materials ordering problems lend themselves particularly well to computer based systems to insure the consistency and completeness of the purchasing process. In the manufacturing realm, the use of automated *materials requirements planning* systems is common. In these systems, the master production schedule, inventory records and product component lists are merged to determine what items must be ordered, when they should be ordered, and how much of each item should be ordered in each time period. The heart of these calculations is simple arithmetic: the projected demand for each material item in each period is subtracted from the available inventory. When the inventory becomes too low, a new order is recommended. For items that are non-standard or not kept in inventory, the calculation is even simpler since no inventory must be considered. With a materials requirement system, much of the detailed record keeping is automated and project managers are alerted to purchasing requirements.

BENEFITS FOR MATERIALS MANAGEMENT SYSTEMS

From a study of twenty heavy construction sites, the following benefits from the introduction of materials management systems were noted:

- In one project, a 6% reduction in craft labor costs occurred due to the improved availability of materials as needed on site. On other projects, an 8% savings due to reduced delay for materials was estimated.
- A comparison of two projects with and without a materials management system revealed a change in productivity from 1.92 man-hours per unit without a system to 1.14 man-hours per unit with a new system. Again, much of this difference can be attributed to the timely availability of materials.
- Warehouse costs were found to decrease 50% on one project with the introduction of improved inventory management, representing a savings of $ 92,000. Interest charges for inventory also declined, with one project reporting a cash flow savings of $ 85,000 from improved materials management.

Against these various benefits, the costs of acquiring and maintaining a materials management system has to be compared. However, management studies suggest that investment in such systems can be quite beneficial.

MATERIAL PROCUREMENT AND DELIVERY

The main sources of information for feedback and control of material procurement are requisitions, bids and quotations, purchase orders and subcontracts, shipping and receiving documents, and invoices. For projects involving the large scale use of critical resources, the owner may initiate the procurement procedure even before the selection of a constructor in order to avoid shortages and delays. Under ordinary circumstances, the constructor will handle the procurement to shop for materials with the best price/performance characteristics specified by the designer. Some overlapping and rehandling in the procurement process is unavoidable, but it should be minimized to insure timely delivery of the materials in good condition.

The materials for delivery to and from a construction site may be broadly classified as:

- Bulk materials,
- Standard off-the-shelf materials, and
- Fabricated members or units.

The process of delivery, including transportation, field storage and installation will be different for these classes of materials. The equipment needed to handle and haul these classes of materials will also be different.

Bulk materials refer to materials in their natural or semi-processed state, such as earthwork to be excavated, wet concrete mix, etc. which are usually encountered in large quantities in construction. Some bulk materials such as earthwork or gravels may be measured in bank (solid in situ) volume. Obviously, the quantities of materials for delivery may be substantially different when expressed in different measures of volume, depending on the characteristics of such materials.

Standard piping and valves are typical examples of standard off-the-shelf materials which are used extensively in the chemical processing industry. Since standard off-the-shelf materials can easily be stockpiled, the delivery process is relatively simple. Fabricated members such as steel beams and columns for buildings are pre-processed in a shop to simplify the field erection procedures. Welded or bolted connections are attached partially to the members which are cut to precise dimensions for adequate fit. Similarly, steel tanks and pressure vessels are often partly or fully fabricated before shipping to the field. In general, if the work can be done in the shop where working conditions can better be controlled, it is advisable to do so, provided that the fabricated members or units can be shipped to the construction site in a satisfactory manner at a reasonable cost.

As a further step to simplify field assembly, an entire wall panel including plumbing and wiring or even an entire room may be prefabricated and shipped to the site. While the field labor is greatly reduced in such cases, "materials" for delivery are in fact manufactured products with value added by another type of labor. With modern means of transporting construction materials and fabricated units, the percentages of costs on direct labor and materials for a project may change if more prefabricated units are introduced in the construction process.

In the construction industry, materials used by a specific craft are generally handled by craftsmen, not by general labor. Thus, electricians handle electrical materials, pipefitters handle pipe materials, etc. This multiple handling diverts scarce skilled craftsmen and contractor supervision into activities which do not directly contribute to construction. Since contractors are not normally in the freight business, they do not perform the tasks of freight delivery efficiently. All these factors tend to exacerbate the problems of freight delivery for very large projects.

FREIGHT DELIVERY FOR THE ALASKA PIPELINE PROJECT

The freight delivery system for the Alaska pipeline project was set up to handle 600,000 tons of materials and supplies. This tonnage did not include the pipes which comprised another 500,000 tons and were shipped through a different routing system. The complexity of this delivery system is illustrated in Figure 6.2. The rectangular boxes denote geographical locations. The points of origin represent plants and factories throughout the US and elsewhere. Some of the materials went to a primary staging point in Seattle and some went directly to Alaska. There were five ports of entry: Valdez, Anchorage, Whittier, Seward and Prudhoe Bay. There was a secondary staging area in Fairbanks and the pipeline itself was divided into six sections. Beyond the Yukon River, there was nothing available but a dirt road for hauling. The amounts of freight in thousands of tons shipped to and from various locations are indicated by the numbers near the network branches (with arrows showing the directions of material flows) and the modes of transportation are noted above the branches. In each of the locations, the contractor had supervision and construction labor to identify materials, unload from transport, determine where the material was going, repackage if required to split shipments, and then re-load material on outgoing transport.

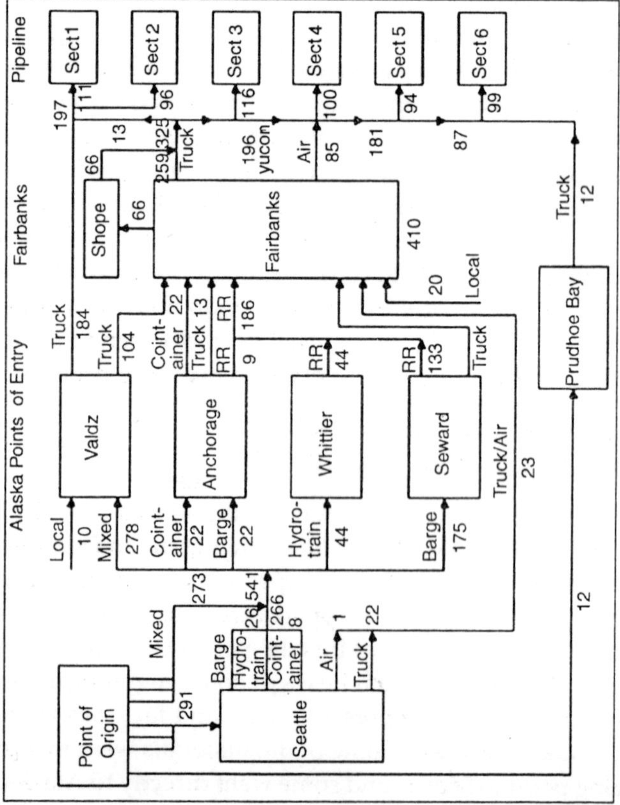

Fig. 6.2 Freight Delivery for the Alaska Pipeline Project

PROCESS PLANT EQUIPMENT PROCUREMENT

The procurement and delivery of bulk materials items such as piping electrical and structural elements involves a series of activities if such items are not standard and/or in stock.

The times required for various activities in the procurement of such items might be estimated to be as follows:

Activities	Duration (days)	Cumulative Duration
Requisition ready by designer	0	0
Owner approval	5	5
Inquiry issued to vendors	3	8
Vendor quotations received	15	23
Complete bid evaluation by designer	7	30
Owner approval	5	35
Place purchase order	5	40
Receive preliminary shop drawings	10	50
Receive final design drawings	10	60
Fabrication and delivery	60-200	120-260

As a result, this type of equipment procurement will typically require four to nine months. Slippage or contraction in this standard schedule is also possible, based on such factors as the extent to which a fabricator is busy.

INVENTORY CONTROL

Once goods are purchased, they represent an *inventory* used during the construction process.

The general objective of inventory control is to minimize the total cost of keeping the inventory while making tradeoffs among the major categories of costs:

- Purchase costs,
- Order cost,
- Holding costs, and
- Unavailable cost.

These cost categories are interrelated since reducing cost

in one category may increase cost in others. The costs in all categories generally are subject to considerable uncertainty.

PURCHASE COSTS

The *purchase cost* of an item is the unit purchase price from an external source including transportation and freight costs. For construction materials, it is common to receive discounts for bulk purchases, so the unit purchase cost declines as quantity increases. These reductions may reflect manufacturers' marketing policies, economies of scale in the material production, or scale economies in transportation. There are also advantages in having homogeneous materials. For example, a bulk order to insure the same colour or size of items such as bricks may be desirable. Accordingly, it is usually desirable to make a limited number of large purchases for materials. In some cases, organizations may consolidate small orders from a number of different projects to capture such bulk discounts; this is a basic saving to be derived from a central purchasing office.

The cost of materials is based on prices obtained through effective bargaining. Unit prices of materials depend on bargaining leverage, quantities and delivery time. Organizations with potential for long-term purchase volume can command better bargaining leverage. While orders in large quantities may result in lower unit prices, they may also increase holding costs and thus cause problems in cash flow. Requirements of short delivery time can also adversely affect unit prices. Furthermore, design characteristics which include items of odd sizes or shapes should be avoided. Since such items normally are not available in the standard stockpile, purchasing them causes higher prices.

The transportation costs are affected by shipment sizes and other factors. Shipment by the full load of a carrier often reduces prices and assures quicker delivery, as the carrier can travel from the origin to the destination of the full load without having to stop for delivering part of the cargo at other stations. Avoiding transshipment is another consideration in reducing shipping cost. While the reduction in shipping costs is a major objective, the requirements of delicate handling of some items may favour a more expensive mode of transportation to avoid breakage and replacement costs.

ORDER COST

The *order* cost reflects the administrative expense of issuing a purchase order to an outside supplier. Order costs include expenses of making requisitions, analyzing alternative vendors, writing purchase orders, receiving materials, inspecting materials, checking on orders, and maintaining records of the entire process. Order costs are usually only a small portion of total costs for material management in construction projects, although ordering may require substantial time.

HOLDING COSTS

The *holding costs* or *carrying costs* are primarily the result of capital costs, handling, storage, obsolescence, shrinkage and deterioration. Capital cost results from the opportunity cost or financial expense of capital tied up in inventory. Once payment for goods is made, borrowing costs are incurred or capital must be diverted from other productive uses. Consequently, a capital carrying cost is incurred equal to the value of the inventory during a period multiplied by the interest rate obtainable or paid during that period. Note that capital costs only accumulate when payment for materials actually occurs; many organizations attempt to delay payments as long as possible to minimize such costs.

Handling and storage represent the movement and protection charges incurred for materials. Storage costs also include the disruption caused to other project activities by large inventories of materials that get in the way. Obsolescence is the risk that an item will lose value because of changes in specifications. Shrinkage is the decrease in inventory over time due to theft or loss. Deterioration reflects a change in material quality due to age or environmental degradation. Many of these *holding cost* components are difficult to predict in advance; a project manager knows only that there is some chance that specific categories of cost will occur. In addition to these major categories of cost, there may be ancillary costs of additional insurance, taxes (many states treat inventories as taxable property), or additional fire hazards. As a general rule, holding costs will typically represent 20 to 40% of the average inventory value over the course of a year; thus if the average material inventory on a project is $ 1 million over a year, the holding cost might be expected to be $200,000 to $400,000.

UNAVAILABILITY COST

The *unavailability cost* is incurred when a desired material is not available at the desired time. In manufacturing industries, this cost is often called the *stockout* or*depletion* cost. Shortages may delay work, thereby wasting labor resources or delaying the completion of the entire project. Again, it may be difficult to forecast in advance exactly when an item may be required or when an shipment will be received. While the project schedule gives one estimate, deviations from the schedule may occur during construction. Moreover, the cost associated with a shortage may also be difficult to assess; if the material used for one activity is not available, it may be possible to assign workers to other activities and, depending upon which activities are critical, the project may not be delayed.

TRADEOFFS OF COSTS IN MATERIALS MANAGEMENT

To illustrate the type of trade-offs encountered in materials management, suppose that a particular item is to be ordered for a project. The amount of time required for processing the order and shipping the item is uncertain. Consequently, the project manager must decide how much lead time to provide in ordering the item. Ordering early and thereby providing a long lead time will increase the chance that the item is available when needed, but it increases the costs of inventory and the chance of spoilage on site.

Let T be the time for the delivery of a particular item, R be the time required for process the order, and S be the shipping time. Then, the minimum amount of time for the delivery of the item is T = R + S. In general, both R and S are random variables; hence T is also a random variable. For the sake of simplicity, we shall consider only the case of instant processing for an order, *i.e.* R = 0. Then, the delivery time T equals the shipping time S.

Since T is a random variable, the chance that an item will be delivered on day t is represented by the probability p(t).

Then, the probability that the item will be delivered on or before t day is given by:

$$P_r\{T \leq t\} = \sum_{u=0}^{t} p(u)$$

If a and b are the lower and upper bounds of possible delivery dates, the expected delivery time is then given by:

$$E[T] = \sum_{t=a}^{b} t[p(t)]$$

The lead time L for ordering an item is the time period ahead of the delivery time, and will depend on the tradeoff between holding costs and unavailability costs. A project manager may want to avoid the unavailable cost by requiring delivery on the scheduled date of use, or may be to lower the holding cost by adopting a more flexible lead time based on the expected delivery time. For example, the manager may make the tradeoff by specifying the lead time to be D days more than the expected delivery time, *i.e.,*

$$L = E[T] + D$$

where D may vary from 0 to the number of additional days required to produce certain delivery on the desired date.

In a more realistic situation, the project manager would also contend with the uncertainty of exactly when the item might be required. Even if the item is *scheduled* for use on a particular date, the work progress might vary so that the desired date would differ. In many cases, greater than expected work progress may result in no savings because materials for future activities are unavailable.

LEAD TIME FOR ORDERING WITH NO PROCESSING TIME

Table summarizes the probability of different delivery times for an item. In this table, the first column lists the possible shipping times (ranging from 10 to 16 days), the second column lists the probability or chance that this shipping time will occur and the third column summarizes the chance that the item arrives on or before a particular date. This table can be used to indicate the chance that the item will arrive on a desired date for different lead times. For example, if the order is placed 12 days in advance of the desired date (so the lead time is 12 days), then there is a 15% chance that the item will arrive exactly on the desired day and a 35% chance that the item will arrive on or before the desired date. Note that this implies that there is a 1-0.35 = 0.65 or 65% chance that the item will not arrive by the desired date with a lead time of 12 days. Given the information in Table, when should the item order be placed?

Table. Delivery Date on Orders and Probability of Delivery for an Example

Delivery Date t	Probability of delivery on date t p(t)	Cummulative probability of delivery by day t Pr{T ≤ t}
10	0.10	0.10
11	0.10	0.20
12	0.15	0.35
13	0.20	0.55
14	0.30	0.85
15	0.10	0.95
16	0.05	1.00

Suppose that the scheduled date of use for the item is in 16 days. To be completely certain to have delivery by the desired day, the order should be placed 16 days in advance. However, the expected delivery date with a 16 day lead time would be:

$$E[T] = \sum_{t=10}^{16} t[p(t)]$$

$$= (10)(0.1) + (11)(0.1) + (12)(0.15) + (13)(0.20) + (14)(0.30) + (15)(0.10) + (16)(0.05) = 13.0$$

Thus, the actual delivery date may be 16-13 = 3 days early, and this early delivery might involve significant holding costs. A project manager might then decide to provide a lead time so that the *expected* delivery date was equal to the desired assembly date as long as the availability of the item was not critical. Alternatively, the project manager might negotiate a more certain delivery date from the supplier.

CONSTRUCTION EQUIPMENT

The selection of the appropriate type and size of construction equipment often affects the required amount of time and effort and thus the job-site productivity of a project. It is therefore important for site managers and construction planners to be familiar with the characteristics of the major types of equipment most commonly used in construction.

EXCAVATION AND LOADING

One family of construction machines used for excavation is broadly classified as a *crane-shovel* as indicated by the variety of machines in Figure 6.3.

The crane-shovel consists of three major components:

- A carrier or mounting which provides mobility and stability for the machine.
- A revolving deck or turntable which contains the power and control units.
- A front end attachment which serves the special functions in an operation.

The type of mounting for all machines in Figure 6.3 is referred to as *crawler mounting*, which is particularly suitable for crawling over relatively rugged surfaces at a job site. Other types of mounting include *truck mounting* and *wheel mounting* which provide greater mobility between job sites, but require better surfaces for their operation. The revolving deck includes a cab to house the person operating the mounting and/or the revolving deck. The types of front end attachments in Figure

6.3 might include a crane with hook, claim shell, dragline, backhoe, shovel and piledriver.

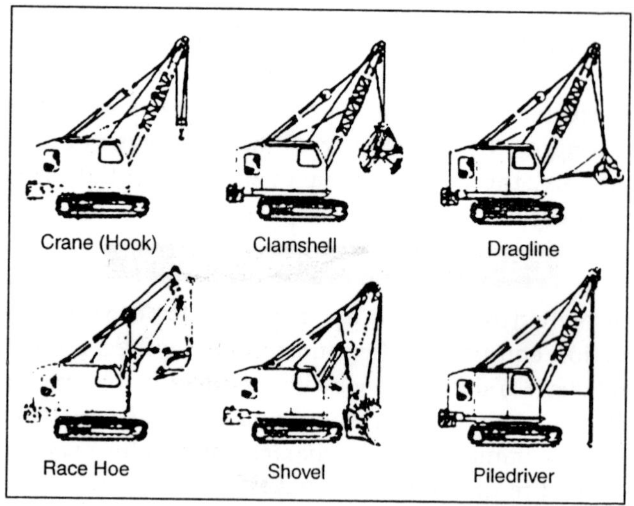

Fig. 6.3 Typical Machines in the Crane-Shovel Family

A tractor consists of a crawler mounting and a non-revolving cab. When an earth moving blade is attached to the front end of a tractor, the assembly is called a bulldozer. When a bucket is attached to its front end, the assembly is known as a loader or bucket loader. There are different types of loaders designed to handle most efficiently materials of different weights and moisture contents.

Scrapers are multiple-units of tractor-truck and blade-bucket assemblies with various combinations to facilitate the loading and hauling of earthwork. Major types of scrapers include single engine two-axle or three axle scrapers, twin-engine all-wheel-drive scrapers, elevating scrapers, and push-pull scrapers. Each type has different characteristics of rolling resistance, maneuverability stability, and speed in operation.

COMPACTION AND GRADING

The function of compaction equipment is to produce higher density in soil mechanically. The basic forces used in compaction are static weight, kneading, impact and vibration. The degree of compaction that may be achieved depends on the properties of soil, its moisture content, the thickness of the soil layer for compaction and the method of compaction.

Some major types of compaction equipment are shown in Figure 6.4, which includes rollers with different operating characteristics.

The function of grading equipment is to bring the earthwork to the desired shape and elevation. Major types of grading equipment include motor graders and grade trimmers. The former is an all-purpose machine for grading and surface finishing, while the latter is used for heavy construction because of its higher operating speed.

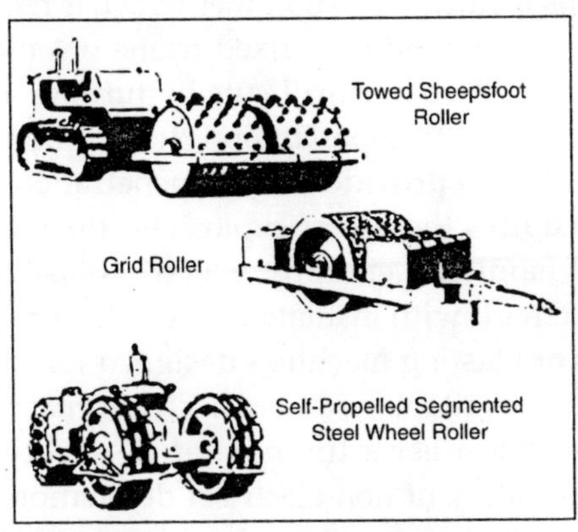

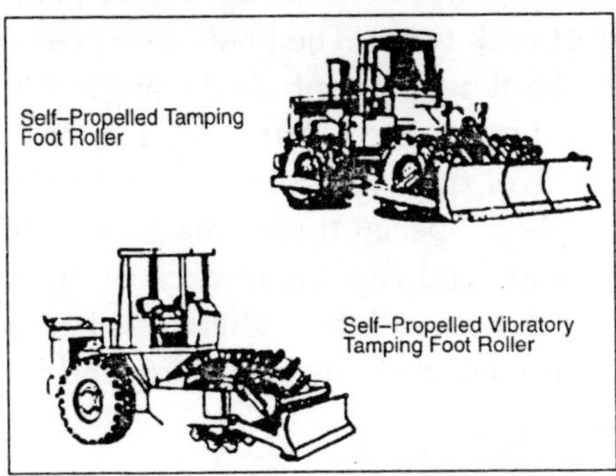

Fig. 6.4 Some Major Types of Compaction Equipment

DRILLING AND BLASTING

Rock excavation is an audacious task requiring special equipment and methods. The degree of difficulty depends on physical characteristics of the rock type to be excavated, such as grain size, planes of weakness, weathering, brittleness and hardness. The task of rock excavation includes loosening, loading, hauling and compacting. The loosening operation is specialized for rock excavation and is performed by drilling, blasting or ripping. Major types of drilling equipment are percussion drills, rotary drills, and rotary-percussion drills. A percussion drill penetrates and cuts rock by impact while it rotates without cutting on the upstroke. Common types of percussion drills include a jackhammer which is hand-held and others which are mounted on a fixed frame or on a wagon or crawl for mobility. A rotary drill cuts by turning a bit against the rock surface. A rotary-percussion drill combines the two cutting movements to provide a faster penetration in rock.

Blasting requires the use of explosives, the most common of which is dynamite. Generally, electric blasting caps are connected in a circuit with insulated wires. Power sources may be power lines or blasting machines designed for firing electric cap circuits. Also available are non-electrical blasting systems which combine the precise timing and flexibility of electric blasting and the safety of non-electrical detonation.

Tractor-mounted rippers are capable of penetrating and prying loose most rock types. The blade or ripper is connected to an adjustable shank which controls the angle at the tip of the blade as it is raised or lowered. Automated ripper control may be installed to control ripping depth and tip angle.

In rock tunneling, special tunnel machines equipped with multiple cutter heads and capable of excavating full diameter of the tunnel are now available. Their use has increasingly replaced the traditional methods of drilling and blasting.

LIFTING AND ERECTING

Derricks are commonly used to lift equipment of materials in industrial or building construction. A derrick consists of a vertical mast and an inclined boom sprouting from the foot of the mast. The mast is held in position by guys or stifflegs connected to a base while a topping lift links the top of the mast and the top of the inclined boom.

A hook in the road line hanging from the top of the inclined boom is used to lift loads. Guy derricks may easily be moved from one floor to the next in a building under construction while stiffleg derricks may be mounted on tracks for movement within a work area.

Tower cranes are used to lift loads to great heights and to facilitate the erection of steel building frames. Horizon boom type tower cranes are most common in highrise building construction. Inclined boom type tower cranes are also used for erecting steel structures.

MIXING AND PAVING

Basic types of equipment for paving include machines for dispensing concrete and bituminous materials for pavement surfaces. Concrete mixers may also be used to mix portland cement, sand, gravel and water in batches for other types of construction other than paving.

A truck mixer refers to a concrete mixer mounted on a truck which is capable of transporting ready mixed concrete from a central batch plant to construction sites. A paving mixer is a self propelled concrete mixer equipped with a boom and a bucket to place concrete at any desired point within a roadway. It can be used as a stationary mixer or used to supply slipform pavers that are capable of spreading, consolidating and finishing a concrete slab without the use of forms.

A bituminous distributor is a truck-mounted plant for generating liquid bituminous materials and applying them to road surfaces through a spray bar connected to the end of the truck. Bituminous materials include both asphalt and tar which have similar properties except that tar is not soluble in petroleum products. While asphalt is most frequently used for road surfacing, tar is used when the pavement is likely to be heavily exposed to petroleum spills.

CONSTRUCTION TOOLS AND OTHER EQUIPMENT

Air compressors and pumps are widely used as the power sources for construction tools and equipment. Common pneumatic construction tools include drills, hammers, grinders, saws, wrenches, staple guns, sandblasting guns, and concrete vibrators. Pumps are used to supply water or to dewater at construction sites and to provide water jets for some types of construction.

AUTOMATION OF EQUIPMENT

The introduction of new mechanized equipment in construction has had a profound effect on the cost and productivity of construction as well as the methods used for construction itself. An exciting example of innovation in this regard is the introduction of computer microprocessors on tools and equipment. As a result, the performance and activity of equipment can be continually monitored and adjusted for improvement. In many cases, automation of at least part of the construction process is possible and desirable. For example, wrenches that automatically monitor the elongation of bolts and the applied torque can be programmed to achieve the best bolt tightness. On grading projects, laser controlled scrapers can produce desired cuts faster and more precisely than wholly manual methods.

EXAMPLE 4-8: TUNNELING EQUIPMENT

In the mid-1980's, some Japanese firms were successful in obtaining construction contracts for tunneling in the United States by using new equipment and methods. For example, the Japanese firm of Ohbayashi won the sewer contract in San Francisco because of its advanced tunneling technology. When a tunnel is dug through soft earth, as in San Francisco, it must be maintained at a few atmospheres of pressure to keep it from caving in. Workers must spend several hours in a pressure chamber before entering the tunnel and several more in decompression afterwards. They can stay inside for only three or four hours, always at considerable risk from cave-ins and asphyxiation. Ohbayashi used the new Japanese "earth-pressure-balance" method, which eliminates these problems.

Whirling blades advance slowly, cutting the tunnel. The loose earth temporarily remains behind to balance the pressure of the compact earth on all sides. Meanwhile, prefabricated concrete segments are inserted and joined with waterproof seals to line the tunnel. Then the loose earth is conveyed away. This new tunneling method enabled Ohbayashi to bid $5 million below the engineer's estimate for a San Francisco sewer. The firm completed the tunnel three months ahead of schedule. In effect, an innovation involving new technology and method led to considerable cost and time savings.

CHOICE OF EQUIPMENT AND STANDARD PRODUCTION RATES

Typically, construction equipment is used to perform essentially repetitive operations, and can be broadly classified according to two basic functions:

- Operators such as cranes, graders, etc. Which stay within the confines of the construction site, and
- Haulers such as dump trucks, ready mixed concrete truck, etc. Which transport materials to and from the site.

In both cases, the cycle of a piece of equipment is a sequence of tasks which is repeated to produce a unit of output. For example, the sequence of tasks for a crane might be to fit and install a wall panel (or a package of eight wall panels) on the side of a building; similarly, the sequence of tasks of a ready mixed concrete truck might be to load, haul and unload two cubic yards (or one truck load) of fresh concrete.

In order to increase job-site productivity, it is beneficial to select equipment with proper characteristics and a size most suitable for the work conditions at a construction site.

In excavation for building construction, for examples, factors that could affect the selection of excavators include:

- *Size of the job*: Larger volumes of excavation will require larger excavators, or smaller excavators in greater number.

- *Activity time constraints*: Shortage of time for excavation may force contractors to increase the size or numbers of equipment for activities related to excavation.
- *Availability of equipment*: Productivity of excavation activities will diminish if the equipment used to perform them is available but not the most adequate.
- *Cost of transportation of equipment*: This cost depends on the size of the job, the distance of transportation, and the means of transportation.
- *Type of excavation*: Principal types of excavation in building projects are cut and/or fill, excavation massive, and excavation for the elements of foundation. The most adequate equipment to perform one of these activities is not the most adequate to perform the others.
- *Soil characteristics*: The type and condition of the soil is important when choosing the most adequate equipment since each piece of equipment has different outputs for different soils. Moreover, one excavation pit could have different soils at different stratums.
- *Geometric characteristics of elements to be excavated*: Functional characteristics of different types of equipment makes such considerations necessary.
- *Space constraints*: The performance of equipment is influenced by the spatial limitations for the movement of excavators.
- *Characteristics of haul units*: The size of an excavator will depend on the haul units if there is a constraint on the size and/or number of these units.
- *Location of dumping areas*: The distance between the construction site and dumping areas could be relevant not only for selecting the type and number of haulers, but also the type of excavators.
- *Weather and temperature*: Rain, snow and severe temperature conditions affect the job-site productivity of labor and equipment.

By comparing various types of machines for excavation,

for example, power shovels are generally found to be the most suitable for excavating from a level surface and for attacking an existing digging surface or one created by the power shovel; furthermore, they have the capability of placing the excavated material directly onto the haulers. Another alternative is to use bulldozers for excavation.

The choice of the type and size of haulers is based on the consideration that the number of haulers selected must be capable of disposing of the excavated materials expeditiously.

Factors which affect this selection include:

- *Output of excavators*: The size and characteristics of the excavators selected will determine the output volume excavated per day.
- *Distance to dump site*: Sometimes part of the excavated materials may be piled up in a corner at the job-site for use as backfill.
- *Probable average speed*: The average speed of the haulers to and from the dumping site will determine the cycle time for each hauling trip.
- *Volume of excavated materials*: The volume of excavated materials including the part to be piled up should be hauled away as soon as possible.
- *Spatial and weight constraints*: The size and weight of the haulers must be feasible at the job site and over the route from the construction site to the dumping area.

Dump trucks are usually used as haulers for excavated materials as they can move freely with relatively high speeds on city streets as well as on highways.

The cycle capacity C of a piece of equipment is defined as the number of output units per cycle of operation under standard work conditions. The capacity is a function of the output units used in the measurement as well as the size of the equipment and the material to be processed. The cycle time T refers to units of time per cycle of operation. The standard production rate R of a piece of construction equipment is defined as the number of output units per unit time.

Hence:

$$R = \frac{C}{T}$$

or,

$$T = \frac{C}{R}$$

The daily standard production rate P_e of an excavator can be obtained by multiplying its standard production rate R_e by the number of operating hours H_e per day.

Thus:

$$P_e = R_e H_e = \frac{C_e H_e}{T_e}$$

where C_e and T_e are cycle capacity (in units of volume) and cycle time (in hours) of the excavator respectively. In determining the daily standard production rate of a hauler, it is necessary to determine first the cycle time from the distance D to a dump site and the average speed S of the hauler. Let T_t be the travel time for the round trip to the dump site, T_o be the loading time and T_d be the dumping time.

Then the travel time for the round trip is given by:

$$T_t = \frac{2D}{S}$$

The loading time is related to the cycle time of the excavator T_e and the relative capacities C_h and C_e of the hauler and the excavator respectively.

In the optimum or standard case:

$$T_o = T_e \frac{C_h}{C_e}$$

For a given dumping time T_d, the cycle time T_h of the hauler is given by:

$$T_h = \frac{2D}{S} + T_e \frac{C_h}{C_e} + T_d$$

The daily standard production rate P_h of a hauler can be obtained by multiplying its standard production rate R_h by the number of operating hours H_h per day.

Hence:

$$P_h = R_h H_h = \frac{C_h H_h}{T_h}$$

This expression assumes that haulers begin loading as soon as they return from the dump site. The number of haulers required is also of interest. Let w denote the swell factor of the soil such that wP_e denotes the daily volume of loose excavated materials resulting from the excavation volume P_e.

Then the approximate number of haulers required to dispose of the excavated materials is given by:

$$N_h = \frac{wP_e}{P_h}$$

While the standard production rate of a piece of equipment is based on "standard" or ideal conditions, equipment productivities at job sites are influenced by actual work conditions and a variety of inefficiencies and work stoppages. As one example, various factor adjustments can be used to account in a approximate fashion for actual site conditions. If the conditions that lower the standard production rate are denoted by n factors F_1, F_2..., F_n, each of which is smaller than 1, then the actual equipment productivity R′ at the job site can be related to the standard production rate R as follows:

$$R' \approx RF_1F_2...F_n$$

On the other hand, the cycle time T′ at the job site will be increased by these factors, reflecting actual work conditions.

If only these factors are involved, T′ is related to the standard cycle time T as:

$$T' \approx \frac{T}{F_1F_2...F_n}$$

Each of these various adjustment factors must be

determined from experience or observation of job sites. For example, a bulk composition factor is derived for bulk excavation in building construction because the standard production rate for general bulk excavation is reduced when an excavator is used to create a ramp to reach the bottom of the bulk and to open up a space in the bulk to accommodate the hauler. In addition to the problem of estimating the various factors, F_1, F_2..., F_n, it may also be important to account for interactions among the factors and the exact influence of particular site characteristics.

Daily Standard Production Rate of a Power Shovel

A power shovel with a dipper of one cubic yard capacity has a standard operating cycle time of 30 seconds. Find the daily standard production rate of the shovel.

For $C_e = 1$ cu. yd., $T_e = 30$ sec. and $H_e = 8$ hours, the daily standard production rate is found from Equation as follows:

$$P_e = \frac{(1\text{yd}^3)(8\text{hr})(3{,}600\,\text{sec/hr})}{30\,\text{sec}} = 960\text{yd}^3$$

In practice, of course, this standard rate would be modified to reflect various production inefficiencies, as described in Example.

Daily Standard Production Rate of a Dump Truck

A dump truck with a capacity of 6 cubic yards is used to dispose of excavated materials at a dump site 4 miles away. The average speed of the dump truck is 30 mph and the dumping time is 30 seconds. Find the daily standard production rate of the truck. If a fleet of dump trucks of this capacity is used to dispose of the excavated materials in Example 4-9 for 8 hours per day, determine the number of trucks needed daily, assuming a swell factor of 1.1 for the soil.

The daily standard production rate of a dump truck can be obtained by using Equations:

$$T_t = \frac{(2)(4\text{mi})(3{,}600\,\text{sec/hr})}{(30\text{mi/hr})} = 960\,\text{sec}$$

$$T_o = (30\,\text{sec})\left[\frac{6\text{yd}^3}{1\text{yd}^3}\right] = 180\,\text{sec}$$

$$T_h = 960 + 180 + 30 = 1{,}170\,\text{sec}$$

Hence, the daily hauler productivity is:

$$P_h = \frac{(6\text{yd}^3)(8\text{hr})(3{,}600\,\text{sec/hr})}{(1{,}170\,\text{sec})} = 147.7\text{yd}^3$$

Finally, from Equation the number of trucks required is:

$$N_h = \frac{(1.1)(960\text{yd}^3)}{147.7\text{yd}^3} = 7.1$$

implying that 8 trucks should be used.

Job Site Productivity of a Power Shovel

A power shovel with a dipper of one cubic yard capacity has a standard production rate of 960 cubic yards for an 8-hour day. Determine the job site productivity and the actual cycle time of this shovel under the work conditions at the site that affects its productivity as shown below:

Work Conditions at the Site	Factors
Bulk composition	0.954
Soil properties and water content	0.983
Equipment idle time for worker breaks	0.8
Management efficiency	0.7

Using Equation, the job site productivity of the power shovel per day is given by:

$$P'_e = (960\text{yd}^3)(0.954)(0.983)(0.8)(0.7) = 504\text{yd}^3$$

The actual cycle time can be determined as follows:

$$T'_e = \frac{(30\,\text{sec})}{(0.954)(0.983)(0.8)(0.7)} = 57\,\text{sec}$$

Noting Equation, the actual cycle time can also be obtained from the relation $T'_e = (C_e H_e)/P'_e$.

Thus:

$$T'_e = \frac{(1yd^3)(8hr)(3{,}600\,sec/hr)}{504yd^3} = 57\,sec$$

Job Site Productivity of a Dump Truck

A dump truck with a capacity of 6 cubic yards is used to dispose of excavated materials. The distance from the dump site is 4 miles and the average speed of the dump truck is 30 mph. The job site productivity of the power shovel per day is 504 cubic yards, which will be modified by a swell factor of 1.1.

The only factors affecting the job site productivity of the dump truck in addition to those affecting the power shovel are 0.80 for equipment idle time and 0.70 for management efficiency. Determine the job site productivity of the dump truck.

If a fleet of such trucks is used to haul the excavated material, find the number of trucks needed daily.

The actual cycle time T'_h of the dump truck can be obtained by summing the actual times for traveling, loading and dumping:

$$T'_t = \frac{T_t}{F_1F_2}\frac{(2)(4mi)(3{,}600\,sec/hr)}{(30mi/hr)(0.8)(0.7)} = 1{,}714\,sec$$

$$T'_o = \frac{T'_e C_h}{F_1F_2C_e}\left[\frac{(57\,sec)}{(0.8)(0.7)}\right]\left[\frac{6yd^3}{1yd^3}\right] = 611\,sec$$

$$T'_d = \frac{T'_d}{F_1F_2}\left[\frac{(30\,sec)}{(0.8)(0.7)}\right] = 54\,sec$$

Hence, the actual cycle time is:

$$T'_h = T'_t + T'_o + T'_d = 1{,}714 + 611 + 54 = 2{,}379\,sec$$

The jobsite productivity P'_h of the dump truck per day is:

$$P'_h = \frac{C_h H_h}{T'_h} = \frac{(6yd^3)(8hr)(3{,}600\,sec/hr)}{2{,}379\,sec} = 72.6yd^3$$

The number of trucks needed daily is:

$$N'_h = \frac{wP'_e}{P'_h} = \frac{(1.1)(504yd^3)}{72.6yd^3} = 7.6$$

so 8 trucks are required.

CONSTRUCTION PROCESSES

The previous sections described the primary inputs of labor, material and equipment to the construction process. At varying levels of detail, a project manager must insure that these inputs are effectively coordinated to achieve an efficient construction process. This coordination involves both strategic decisions and tactical management in the field. For example, strategic decisions about appropriate technologies or site layout are often made during the process of construction planning. During the course of construction, foremen and site managers will make decisions about work to be undertaken at particular times of the day based upon the availability of the necessary resources of labor, materials and equipment. Without coordination among these necessary inputs, the construction process will be inefficient or stop altogether.

STEEL ERECTION

Erection of structural steel for buildings, bridges or other facilities is an example of a construction process requiring considerable coordination. Fabricated steel pieces must arrive on site in the correct order and quantity for the planned effort during a day. Crews of steelworkers must be available to fit pieces together, bolt joints, and perform any necessary welding. Cranes and crane operators may be required to lift fabricated components into place; other activities on a job site may also be competing for use of particular cranes. Welding equipment, wrenches and other hand tools must be readily available. Finally, ancillary materials such as bolts of the correct size must be provided. In coordinating a process such as steel erection, it is common to assign different tasks to specific crews. For example, one crew may place members in place and insert a few bolts in joints in a specific area. A following crew would be assigned to finish bolting, and a third crew might perform necessary welds or attachment of brackets for items such as curtain walls. With the required coordination among these resources, it is easy to see how poor management or other problems can result in considerable inefficiency. For example, if a shipment of fabricated steel is improperly prepared, the crews and equipment on site may have to wait for new deliveries.

Construction Process Simulation Models

Computer based simulation of construction operations can be a useful although laborious tool in analyzing the efficiency of particular processes or technologies. These tools tend to be either oriented towards modeling resource processes or towards representation of spatial constraints and resource movements. Later chapters will describe simulation in more detail, but a small example of a construction operation model can be described here. The process involved placing concrete within existing formwork for the columns of a new structure.

A crane-and-bucket combination with one cubic yard capacity and a flexible "elephant trunk" was assumed for placement. Concrete was delivered in trucks with a capacity of eight cubic yards. Because of site constraints, only one truck could be moved into the delivery position at a time. Construction workers and electric immersion-type concrete vibrators were also assumed for the process.

The simulation model of this process is illustrated in Figure 6.5. Node 2 signals the availability of a concrete truck arriving from the batch plant. As with other circular nodes in Figure 6.5, the availability of a truck may result in a resource waiting or *queueing* for use. If a truck (node 2) and the crane (node 3) are both available, then the crane can load and hoist a bucket of concrete (node 4). As with other rectangular nodes in the model, this operation will require an appreciable period of time. On the completion of the load and hoist operations, the bucket (node 5) is available for concrete placement. Placement is accomplished by having a worker guide the bucket's elephant trunk between the concrete forms and having a second worker operate the bucket release lever. A third laborer operates a vibrator in the concrete while the bucket (node 8) moves back to receive a new load. Once the concrete placement is complete, the crew becomes available to place a new bucket load (node 7). After two buckets are placed, then the column is complete (node 9) and the equipment and crew can move to the next column (node 10). After the movement to the new column is complete, placement in the new column can begin (node 11). Finally, after a truck is emptied (nodes 12 and 13), the truck departs and a new truck can enter the delivery stall (node 14) if one is waiting.

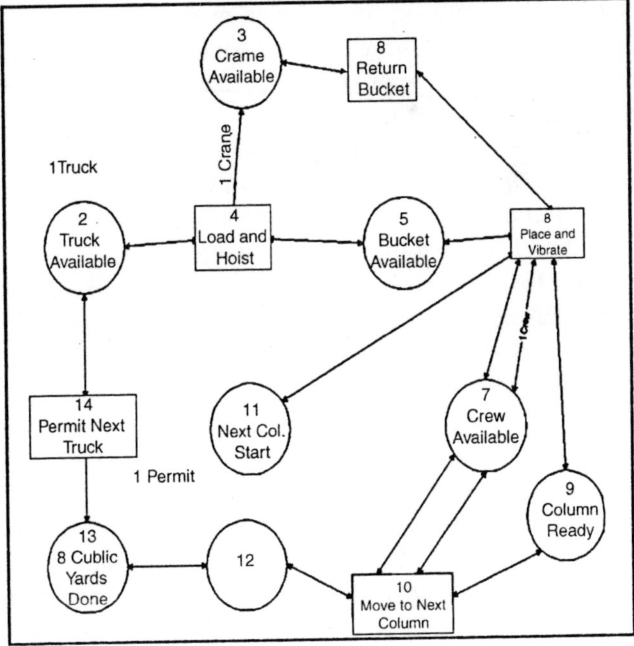

Fig. 6.5 Illustration of a Concrete-Placing Simulation Model

Application of the simulation model consists of tracing through the time required for these various operations. Events are also simulated such as the arrival times of concrete trucks. If random elements are introduced, numerous simulations are required to estimate the actual productivity and resource requirements of the process.

For example, one simulation of this process using four concrete trucks found that a truck was waiting 83% of the time with an average wait at the site of 14 minutes. This type of simulation can be used to estimate the various productivity adjustment factors described in the previous section.

QUEUES AND RESOURCE BOTTLENECKS

A project manager needs to insure that resources required for and/or shared by numerous activities are adequate. Problems in this area can be indicated in part by the existence of queues of resource demands during construction operations.

A *queue* can be a *waiting line* for service. One can imagine a queue as an orderly line of customers waiting for a stationary server such as a ticket seller. However, the demands for service might not be so neatly arranged. For example, we can speak of the *queue* of welds on a building site waiting for inspection. In this case, demands do not come to the server, but a roving inspector travels among the waiting service points.

Waiting for resources such as a particular piece of equipment or a particular individual is an endemic problem on construction sites. If workers spend appreciable portions of time waiting for particular tools, materials or an inspector, costs increase and productivity declines. Insuring adequate resources to serve expected demands is an important problem during construction planning and field management.

In general, there is a trade-off between waiting times and utilization of resources. Utilization is the proportion of time a particular resource is in productive use. Higher amounts of resource utilization will be beneficial as long as it does not impose undue costs on the entire operation. For example, a welding inspector might have one hundred per cent utilization, but workers throughout the jobsite might be wasting inordinate time waiting for inspections. Providing additional inspectors may be cost effective, even if they are not utilized at all times. A few conceptual models of queueing systems may be helpful to construction planners in considering the level of adequate resources to provide.

First, we shall consider the case of time-varying demands and a server with a constant service rate. This might be the situation for an elevator in which large demands for transportation occur during the morning or at a shift change. Second, we shall consider the situation of randomly arriving demands for service and constant service rates. Finally, we shall consider briefly the problems involving multiple serving stations.

SINGLE-SERVER WITH DETERMINISTIC ARRIVALS AND SERVICES

Suppose that the cumulative number of demands for service or "customers" at any time t is known and equal to the value of the function A(t). These "customers" might be crane loads, weld inspections, or any other defined group of items to be serviced.

Suppose further that a single server is available to handle these demands, such as a single crane or a single inspector. For this model of queueing, we assume that the server can handle customers at some constant, maximum rate denoted as x "customers" per unit of time. This is a maximum rate since the server may be idle for periods of time if no customers are waiting. This system is *deterministic* in the sense that both the arrival function and the service process are assumed to have no random or unknown component.

A cumulative arrival function of customers, A(t), is shown in Figure 6.6 in which the vertical axis represents the cumulative number of customers, while the horizontal axis represents the passage of time. The arrival of individual customers to the queue would actually represent a unit step in the arrival function A(t), but these small steps are approximated by a continuous curve in the figure 6.6.

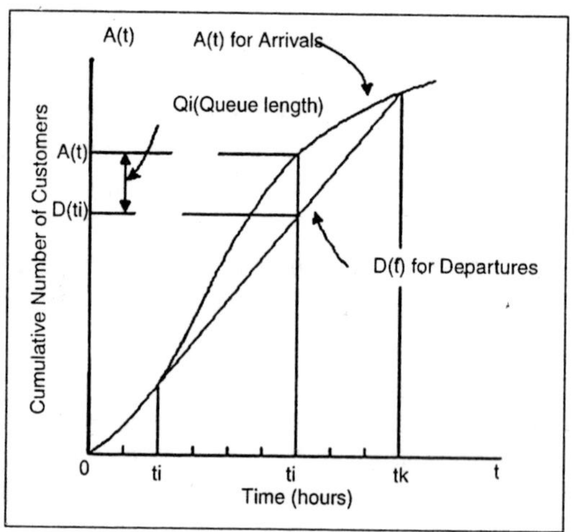

Fig. 6.6 Cumulative Arrivals and Departures in a Deterministic Queue

The rate of arrivals for a unit time interval Δt from t-1 to t is given by:

$$\Delta A_t = A(t) - A(t-1)$$

While an hour or a minute is a natural choice as a unit time interval, other time periods may also be used as long as the passage of time is expressed as multiples of such time periods. For instance, if half an hour is used as unit time interval for a process involving ten hours, then the arrivals should be represented by 20 steps of half hour each.

Hence, the unit time interval between t-1 and t is Δt = t-(t-1) = 1, and the slope of the cumulative arrival function in the interval is given by:

$$D_f = Wh_t + (W - W_t)C_t$$

The cumulative number of customers served over time is represented by the cumulative departure function D(t).

While the maximum service rate is x per unit time, the actual service rate for a unit time interval Δt from t-1 to t is:

$$\Delta D_t = D(t) - D(t-1)$$

The slope of the cumulative departure function is:

$$D'(t) = \frac{D(t) - D(t-1)}{\Delta t} = D(t) - D(t-1)$$

Any time that the rate of arrivals to the queue exceeds the maximum service rate, then a queue begins to form and the cumulative departures will occur at the maximum service rate. The cumulative departures from the queue will proceed at the maximum service rate of x "customers" per unit of time, so that the slope of D(t) is x during this period.

The cumulative departure function D(t) can be readily constructed graphically by running a ruler with a slope of x along the cumulative arrival function A(t). As soon as the function A(t) climbs above the ruler, a queue begins to form. The maximum service rate will continue until the queue disappears, which is represented by the convergence of the cumulative arrival and departure functions A(t) and D(t).

With the cumulative arrivals and cumulative departure functions represented graphically, a variety of service indicators can be readily obtained as shown in Figure 6.6. Let A′(t) and D′(t) denote the derivatives of A(t) and D(t) with respect to t, respectively.

For $0 \le t \le t_i$ in which $A'(t) \le x$, there is no queue. At $t = t_i$, when $A'(t) > D'(t)$, a queue is formed. Then $D'(t) = x$ in the interval $t_i \le t \le t_k$. As A′(t) continues to increase with increasing t, the queue becomes longer since the service rate $D'(t) = x$ cannot catch up with the arrivals. However, when again $A'(t) \le D'(t)$ as t increases, the queue becomes shorter until it reaches 0 at $t = t_k$. At any given time t, the queue length is,

$$Q(t) = A(t) - D(t)$$

For example, suppose a queue begins to form at time t_i and is dispersed by time t_k. The maximum number of customers waiting or queue length is represented by the maximum difference between the cumulative arrival and cumulative departure functions between t_i and t_k, *i.e.* the maximum value of Q(t). The total waiting time for service is indicated by the total area between the cumulative arrival and cumulative departure functions. Generally, the arrival rates ΔA_t = 1, 2..., n periods of a process as well as the maximum service rate x are known.

Then the cumulative arrival function and the cumulative departure function can be constructed systematically together with other pertinent quantities as follows:

- Starting with the initial conditions D(t-1)=0 and Q(t-1)=0 at t=1, find the actual service rate at t=1:

$$\Delta D_1 = \text{minimum}\{x; A_1\}$$

- Starting with A(t-1)=0 at t=1, find the cumulative arrival function for t=2,3...,n accordingly:

$$A(t) = A(t-1) = \Delta A_t$$

- Compute the queue length for t=1,2...,n.

$$A(t) = A(t-1) = \Delta A_t$$

- Compute ΔD_t for t=2,3...,n after Q(t-1) is found first for each t:

$$\Delta D_t = \text{minimum}\{x; Q(t-1) + \Delta A_t\}$$

- If A′(t) > x, find the cumulative departure function in the time period between t_i where a queue is formed and t_k where the queue dissipates:

$$D(t) = D(t-1) + \Delta D_t$$

- Compute the waiting time Δw for the arrivals which are waiting for service in interval Δt:

$$\Delta W = Q(t)(\Delta t)$$

- Compute the total waiting time W over the time period between t_i and t_k.

$$W = \sum_{t=t_i}^{t_k} \Delta w$$

- Compute the average waiting time w for arrivals which are waiting for service in the process.

$$W = \frac{W}{A(t_k) - A(t_i)}$$

This simple, deterministic model has a number of implications for operations planning. First, an increase in the maximum service rate will result in reductions in waiting time and the maximum queue length.

Such increases might be obtained by speeding up the service rate such as introducing shorter inspection procedures or installing faster cranes on a site. Second, altering the pattern of cumulative arrivals can result in changes in total waiting time and in the maximum queue length. In particular, if the maximum arrival rate never exceeds the maximum service rate, no queue will form, or if the arrival rate always exceeds the maximum service rate, the bottleneck cannot be dispersed. Both cases are shown in Figure 6.7.

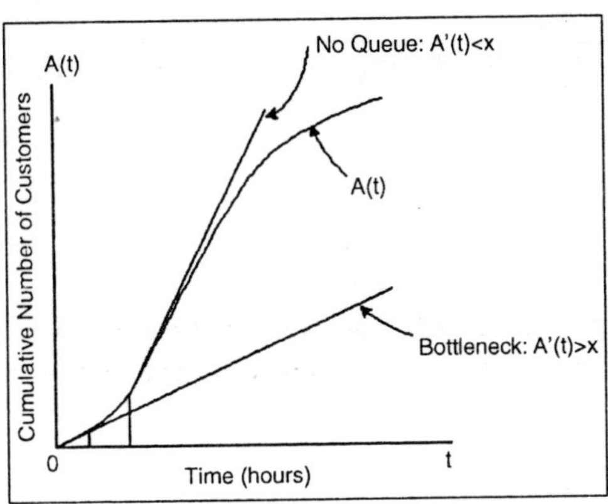

Fig. 6.7 Cases of no Queue and Permanent Bottleneck

A practical means to alter the arrival function and obtain these benefits is to inaugurate a reservation system for customers. Even without drawing a graph such as Figure 6.6, good operations planners should consider the effects of different operation or service rates on the flow of work. Clearly, service rates less than the expected arrival rate of work will result in resource bottlenecks on a job.

SINGLE-SERVER WITH RANDOM ARRIVALS AND CONSTANT SERVICE RATE

Suppose that arrivals of "customers" to a queue are not deterministic or known as in Figure 6.6. In particular, suppose that "customers" such as joints are completed or crane loads arrive at random intervals. What are the implications for the smooth flow of work? Unfortunately, bottlenecks and queues may arise in this situation even if the maximum service rate is larger than the average or expected arrival rate of customers. This occurs because random arrivals will often bunch together, thereby temporarily exceeding the capacity of the system. While the average arrival rate may not change over time, temporary resource shortages can occur in this circumstance. Let w be the average waiting time, a be the average arrival rate of customers, and x be the deterministic constant service rate (in customers per unit of time).

Then, the expected average time for a customer in this situation is given by:

$$W = \frac{a}{2x^2\left[1-\frac{a}{x}\right]}$$

If the average utilization rate of the service is defined as the ratio of the average arrival rate and the constant service rate, *i.e.,*

$$u = \frac{a}{x}$$

Then, Equation becomes:

$$w = \frac{u}{2x(1-u)}$$

In this equation, the ratio u of arrival rate to service rate is very important: if the average arrival rate approaches the service rate, the waiting time can be very long. If a ≥ x, then the queue expands indefinitely. Resource bottlenecks will occur with random arrivals unless a measure of extra service capacity is available to accommodate sudden bunches in the arrival stream. Figure 6.8 illustrates the waiting time resulting from different combinations of arrival rates and service times.

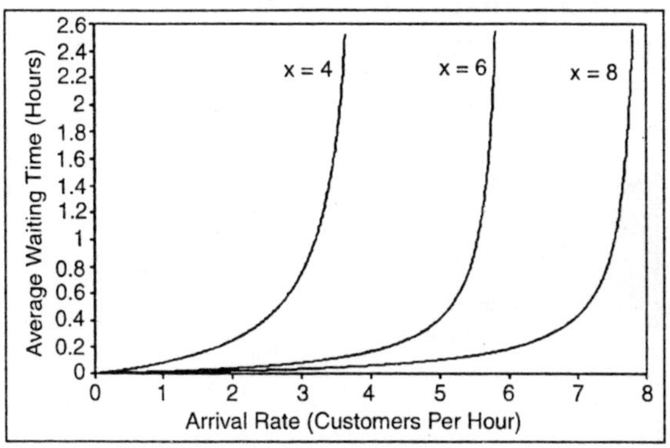

Fig. 6.8 Illustrative Waiting Times for Different Average Arrival Rates and Service Times

MULTIPLE SERVERS

Both of the simple models of service performance described above are limited to single servers. In operations planning, it is commonly the case that numerous operators are available and numerous stages of operations exist. In these circumstances, a planner typically attempts to match the service rates occurring at different stages in the process. For example, construction of a high rise building involves a series of operations on each floor, including erection of structural elements, pouring or assembling a floor, construction of walls, installation of HVAC (Heating, ventilating and air conditioning) equipment, installation of plumbing and electric wiring, etc.

A smooth construction process would have each of these various activities occurring at different floors at the same time without large time gaps between activities on any particular floor. Thus, floors would be installed soon after erection of structural elements, walls would follow subsequently, and so on. From the standpoint of a queueing system, the planning problem is to insure that the productivity or service rate per floor of these different activities are approximately equal, so that one crew is not continually waiting on the completion of a preceding activity or interfering with a following activity. In the realm of manufacturing systems, creating this balance among operations is called *assembly line* balancing.

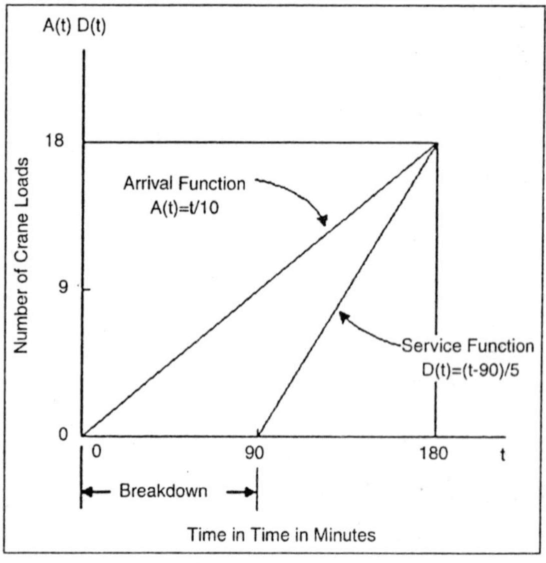

Fig. 6.9 Arrivals and Services of Crane Loads with a Crane Breakdown

Effect of a Crane Breakdown

Suppose that loads for a crane are arriving at a steady rate of one every ten minutes. The crane has the capacity to handle one load every five minutes. Suppose further that the crane breaks down for ninety minutes. How many loads are delayed, what is the total delay, and how long will be required before the crane can catch up with the backlog of loads? The cumulative arrival and service functions are graphed in Figure 6.9. Starting with the breakdown at time zero, nine loads arrive during the ninety minute repair time. From Figure 6.9, an additional nine loads arrive before the entire queue is served. Algebraically, the required time for service, t, can be calculated by noted that the number of arrivals must equal the number of loads served.

Thus:

$$A(t) = \frac{t}{10} \text{ for } t \geq 0$$

$$D_1(t) = 0 \text{ for } 0 \leq t \leq 90 \text{min}$$

$$D_2(t) = \frac{t-90}{5} \text{ for } t \geq 90 \text{ min}$$

A queue is formed at t = 0 because of the breakdown, but it dissipates at A(t) = D_2(t). Let,

$$\frac{t}{10} = \frac{t-90}{5}$$

from which we obtain t = 180 min. Hence,

$$A(180) = D_2(180) = 18 \text{ loads}$$

The total waiting time W can be calculated as the area between the cumulative arrival and service functions.

Algebraically, this is conveniently calculated as the difference in the areas of two triangles:

$$W = \frac{(18)(180)}{2} - \frac{(18)(90)}{2} = 810 \text{ min}$$

so the average delay per load is w = 810/18 = 45 minutes.

Waiting Time with Random Arrivals

Suppose that material loads to be inspected arrive randomly but with an average of 5 arrivals per hour. Each load requires ten minutes for an inspection, so an inspector can handle six loads per hour.

Inspections must be completed before the material can be unloaded from a truck. The cost per hour of holding a material load in waiting is $30, representing the cost of a driver and a truck. In this example, the arrival rate, a, equals 5 arrivals per hour and the service rate, x, equals 6 material loads per hour.

Then, the average waiting time of any material load for u = 5/6 is:

$$\frac{5/6}{(2)(6)-(1-5/6)} = 0.4\text{hr}$$

At a resource cost of $30.00 per hour, this waiting would represent a cost of (30)(0.4)(5) = $60.00 per hour on the project.

In contrast, if the possible service rate is x = 10 material loads per hour, then the expected waiting time of any material load for u = 5/10 = 0.5 is:

$$\frac{0.5}{(2)(10)(1-0.5)} = 0.05\text{hr}$$

which has only a cost of (30)(0.05)(5) = $7.50 per hour.

Delay of Lift Loads on a Building Site

Suppose that a single crane is available on a building site and that each lift requires three minutes including the time for attaching loads. Suppose further that the cumulative arrivals of lift loads at different time periods are as follows:

6:00-7:00 A.M.	4 per hour	12:00-4:00 P.M.	8 per hour
7:00-8:00 A.M.	15 per hour	4:00-6:00 P.M.	4 per hour
8:00-11:00 A.M.	25 per hour	6:00P.M.-6:00 A.M.	0 per hour
11:00-12:00 A.M.	5 per hour		

Using the above information of arrival and service rates:

- Find the cumulative arrivals and cumulative number of loads served as a function of time, beginning with 6:00 AM.
- Estimate the maximum queue length of loads waiting for service. What time does the maximum queue occur?

- Estimate the total waiting time for loads.
- Graph the cumulative arrival and departure functions.

The maximum service rate x = 60 min/3 min per lift = 20 lifts per minute. The detailed computation can be carried out in the Table, and the graph of A(t) and D(t) is given.

Table. Computation of Queue Length and Waiting Time

Period	Arrival rate	Cumulative arrivals *A(T)*	Queue	Departure rate	Cumulative departures *D(T)*	Waiting time
6-7:00	4	4	0	4	4	0
7-8:00	15	19	0	15	19	0
8-9:00	25	44	5	20	39	5
9-10:00	25	69	10	20	59	10
10-11:00	25	94	15	20	79	15
11-12:00	5	99	0	20	99	0
12-1:00	8	107	0	8	107	0
1-2:00	8	115	0	8	115	0
2-3:00	8	123	0	8	123	0
3-4:00	8	131	0	8	131	0
4-5:00	4	135	0	4	135	0
5-6:00	4	139	0	4	139	0
6-7:00	0	139	0	0	139	0
7-8:00	0	139	0	0	139	0
				Total waiting time = 30		
				Maximum queue = 15		

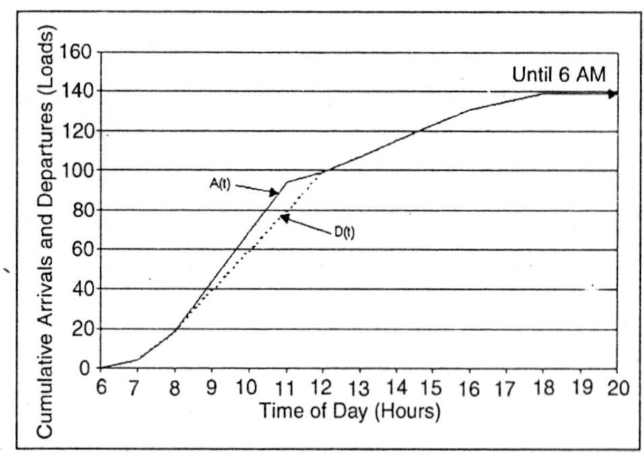

Fig. 6.10 Delay of Lift Loads on a Building Site

7

Construction Planning

BASIC CONCEPTS IN THE DEVELOPMENT OF CONSTRUCTION PLANS

Construction planning is a fundamental and challenging activity in the management and execution of construction projects. It involves the choice of technology, the definition of work tasks, the estimation of the required resources and durations for individual tasks, and the identification of any interactions among the different work tasks. A good construction plan is the basis for developing the budget and the schedule for work. Developing the construction plan is a critical task in the management of construction, even if the plan is not written or otherwise formally recorded. In addition to these technical aspects of construction planning, it may also be necessary to make organizational decisions about the relationships between project participants and even which organizations to include in a project. For example, the extent to which sub-contractors will be used on a project is often determined during construction planning.

Most people, if you describe a train of events to them, will tell you what the result would be. They can put those events together in their minds, and argue from them that something will come to pass. There are few people, however, who, if you told them a result, would be able to evolve from their own inner consciousness what the steps were which led up to that result. This power is what I mean when I talk of reasoning backward.

Like a detective, a planner begins with a result (*i.e.* a facility design) and must synthesize the steps required to yield this result. Essential aspects of construction planning include the *generation* of required activities, *analysis* of the implications of these activities, and *choice* among the various alternative means of performing activities.

In contrast to a detective discovering a single train of events, however, construction planners also face the normative problem of choosing the best among numerous alternative plans. Moreover, a detective is faced with an observable result, whereas a planner must imagine the final facility as described in the plans and specifications.

In developing a construction plan, it is common to adopt a primary emphasis on either cost control or on schedule control as illustrated in Fig 7.1. Some projects are primarily divided into expense categories with associated costs. In these cases, construction planning is cost or expense oriented. Within the categories of expenditure, a distinction is made between costs incurred directly in the performance of an activity and indirectly for the accomplishment of the project. For example, borrowing expenses for project financing and overhead items are commonly treated as indirect costs. For other projects, scheduling of work activities over time is critical and is emphasized in the planning process.

In this case, the planner insures that the proper precedences among activities are maintained and that efficient scheduling of the available resources prevails. Traditional scheduling procedures emphasize the maintenance of task precedences (resulting in *critical path scheduling* procedures) or efficient use of resources over time (resulting in *job shop scheduling* procedures). Finally, most complex projects require consideration of both cost and scheduling over time, so that planning, monitoring and record keeping must consider both dimensions. In these cases, the integration of schedule and budget information is a major concern.

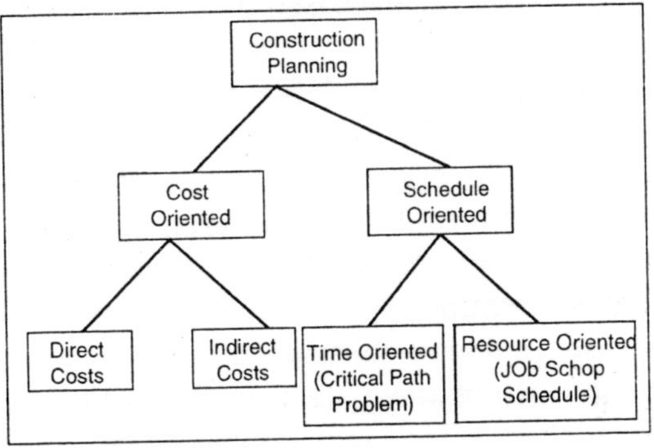

Fig. 7.1 Alternative Emphases in Construction Planning

In this chapter, we shall consider the functional requirements for construction planning such as technology choice, work breakdown, and budgeting. Construction planning is not an activity which is restricted to the period after the award of a contract for construction. It should be an essential activity during the facility design. Also, if problems arise during construction, re-planning is required.

CHOICE OF TECHNOLOGY AND CONSTRUCTION METHOD

As in the development of appropriate alternatives for facility design, choices of appropriate technology and methods for construction are often ill-structured yet critical ingredients in the success of the project. For example, a decision whether to pump or to transport concrete in buckets will directly affect the cost and duration of tasks involved in building construction. A decision between these two alternatives should consider the relative costs, reliabilities, and availability of equipment for the two transport methods. Unfortunately, the exact implications of different methods depend upon numerous considerations for which information may be sketchy during the planning phase, such as the experience and expertise of workers or the particular underground condition at a site. In selecting among alternative methods and technologies, it may be necessary to formulate a number of construction plans based on alternative methods or assumptions.

Once the full plan is available, then the cost, time and reliability impacts of the alternative approaches can be reviewed. This examination of several alternatives is often made explicit in bidding competitions in which several alternative designs may be proposed or *value engineering* for alternative construction methods may be permitted.

In this case, potential constructors may wish to prepare plans for each alternative design using the suggested construction method as well as to prepare plans for alternative construction methods which would be proposed as part of the value engineering process.

In forming a construction plan, a useful approach is to simulate the construction process either in the imagination of the planner or with a formal computer based simulation technique. By observing the result, comparisons among different plans or problems with the existing plan can be identified. For example, a decision to use a particular piece of equipment for an operation immediately leads to the question of whether or not there is sufficient access space for the equipment. Three dimensional geometric models in a computer aided design (CAD) system may be helpful in simulating space requirements for operations and for identifying any interferences. Similarly, problems in resource availability identified during the simulation of the construction process might be effectively forestalled by providing additional resources as part of the construction plan.

A ROADWAY REHABILITATION

An example from a roadway rehabilitation project in Pittsburgh, PA can serve to illustrate the importance of good construction planning and the effect of technology choice. In this project, the decks on overpass bridges as well as the pavement on the highway itself were to be replaced. The initial construction plan was to work outward from each end of the overpass bridges while the highway surface was replaced below the bridges.

As a result, access of equipment and concrete trucks to the overpass bridges was a considerable problem. However, the highway work could be staged so that each overpass bridge was accessible from below at prescribed times. By pumping concrete up to the overpass bridge deck from the highway below, costs were reduced and the work was accomplished much more quickly.

LASER LEVELING

An example of technology choice is the use of laser leveling equipment to improve the productivity of excavation and grading. In these systems, laser surveying equipment is erected on a site so that the relative height of mobile equipment is known exactly. This height measurement is accomplished by flashing a rotating laser light on a level plane across the construction site and observing exactly where the light shines on receptors on mobile equipment such as graders. Since laser light does not disperse appreciably, the height at which the laser shines anywhere on the construction site gives an accurate indication of the height of a receptor on a piece of mobile equipment. In turn, the receptor height can be used to measure the height of a blade, excavator bucket or other piece of equipment. Combined with electro-hydraulic control systems mounted on mobile equipment such as bulldozers, graders and scrapers, the height of excavation and grading blades can be precisely and automatically controlled in these systems.

This automation of blade heights has reduced costs in some cases by over 80% and improved quality in the finished product, as measured by the desired amount of excavation or the extent to which a final grade achieves the desired angle. These systems also permit the use of smaller machines and less skilled operators.

However, the use of these semi-automated systems require investments in the laser surveying equipment as well as modification to equipment to permit electronic feedback control units. Still, laser leveling appears to be an excellent technological choice in many instances.

DEFINING WORK TASKS

At the same time that the choice of technology and general method are considered, a parallel step in the planning process is to define the various work tasks that must be accomplished. These work tasks represent the necessary framework to permit *scheduling* of construction activities, along with estimating the *resources* required by the individual work tasks, and any necessary *precedences* or required sequence among the tasks. The terms work "tasks" or "activities" are often used interchangeably in construction plans to refer to specific, defined items of work. In job shop or manufacturing terminology, a project would be called a "job" and an activity called an "operation", but the sense of the terms is equivalent.

The *scheduling problem* is to determine an appropriate set of activity start time, resource allocations and completion times that will result in completion of the project in a timely and efficient fashion.

Construction planning is the necessary fore-runner to scheduling. In this planning, defining work tasks, technology and construction method is typically done either simultaeously or in a series of iterations.

The definition of appropriate work tasks can be a laborious and tedious process, yet it represents the necessary information for application of formal scheduling procedures. Since construction projects can involve thousands of individual work tasks, this definition phase can also be expensive and time consuming.

Fortunately, many tasks may be repeated in different parts of the facility or past facility construction plans can be used as general models for new projects. For example, the tasks involved in the construction of a building floor may be repeated with only minor differences for each of the floors in the building. Also, standard definitions and nomenclatures for most tasks exist. As a result, the individual planner defining work tasks does not have to approach each facet of the project entirely from scratch.

While repetition of activities in different locations or reproduction of activities from past projects reduces the work involved, there are very few computer aids for the process of defining activities. For the scheduling process itself, numerous computer programmes are available. But for the important task of defining activities, reliance on the skill, judgment and experience of the construction planner is likely to continue.

More formally, an *activity* is any subdivision of project tasks. The set of activities defined for a project should be *comprehensive* or completely *exhaustive* so that all necessary work tasks are included in one or more activities. Typically, each design element in the planned facility will have one or more associated project activities. Execution of an activity requires time and resources, including manpower and equipment, as described in the next section. The time required to perform an activity is called the *duration* of the activity.

The beginning and the end of activities are signposts or *milestones,* indicating the progress of the project. Occasionally, it is useful to define activities which have no duration to mark important events.

For example, receipt of equipment on the construction site may be defined as an activity since other activities would depend upon the equipment availability and the project manager might appreciate formal notice of the arrival. Similarly, receipt of regulatory approvals would also be specially marked in the project plan.

The extent of work involved in any one activity can vary tremendously in construction project plans. Indeed, it is common to begin with fairly coarse definitions of activities and then to further sub-divide tasks as the plan becomes better defined.

As a result, the definition of activities evolves during the preparation of the plan. A result of this process is a natural *hierarchy* of activities with large, abstract functional activities repeatedly sub-divided into more and more specific sub-tasks. For example, the problem of placing concrete on site would have sub-activities associated with placing forms, installing reinforcing steel, pouring concrete, finishing the concrete, removing forms and others. Even more specifically, sub-tasks such as removal and cleaning of forms after concrete placement can be defined. Even further, the sub-task "clean concrete forms" could be subdivided into the various operations:

- Transport forms from on-site storage and unload onto the cleaning station.
- Position forms on the cleaning station.
- Wash forms with water.
- Clean concrete debris from the form's surface.
- Coat the form surface with an oil release agent for the next use.
- Unload the form from the cleaning station and transport to the storage location.

This detailed task breakdown of the activity "clean concrete forms" would not generally be done in standard construction planning, but it is essential in the process of programming or designing a *robot* to undertake this activity since the various specific tasks must be well defined for a robot implementation.

It is generally advantageous to introduce an explicit *hierarchy* of work activities for the purpose of simplifying the presentation and development of a schedule. For example, the initial plan might define a single activity associated with "site clearance."

Later, this single activity might be sub-divided into "re-locating utilities," "removing vegetation," "grading", etc. However, these activities could continue to be identified as sub-activities under the general activity of "site clearance." This hierarchical structure also facilitates the preparation of summary charts and reports in which detailed operations are combined into aggregate or "super"-activities.

More formally, a hierarchical approach to work task definition decomposes the work activity into component parts in the form of a tree. Higher levels in the tree represent decision nodes or summary activities, while branches in the tree lead to smaller components and work activities. A variety of constraints among the various nodes may be defined or imposed, including precedence relationships among different tasks as defined below. Technology choices may be *decomposed* to decisions made at particular nodes in the tree. For example, choices on plumbing technology might be made without reference to choices for other functional activities.

Of course, numerous different activity hierarchies can be defined for each construction plan. For example, upper level activities might be related to facility components such as foundation elements, and then lower level activity divisions into the required construction operations might be made. Alternatively, upper level divisions might represent general types of activities such as electrical work, while lower work divisions represent the application of these operations to specific facility components.

As a third alternative, initial divisions might represent different spatial locations in the planned facility. The choice of a hierarchy depends upon the desired scheme for summarizing work information and on the convenience of the planner. In computerized databases, multiple hierarchies can be stored so that different aggregations or views of the work breakdown structure can be obtained.

The number and detail of the activities in a construction plan is a matter of judgment or convention. Construction plans can easily range between less than a hundred to many thousand defined tasks, depending on the planner's decisions and the scope of the project. If subdivided activities are too refined, the size of the network becomes unwieldy and the cost of planning excessive.

Sub-division yields no benefit if reasonably accurate estimates of activity durations and the required resources cannot be made at the detailed work breakdown level. On the other hand, if the specified activities are too coarse, it is impossible to develop realistic schedules and details of resource requirements during the project.

It is useful to define separate work tasks for:

- Those activities which involve different resources, or
- Those activities which do not require continuous performance.

For example, the activity "prepare and check shop drawings" should be divided into a task for preparation and a task for checking since different individuals are involved in the two tasks and there may be a time lag between preparation and checking.

In practice, the proper level of detail will depend upon the size, importance and difficulty of the project as well as the specific scheduling and accounting procedures which are adopted.

However, it is generally the case that most schedules are prepared with too little detail than too much. It is important to keep in mind that task definition will serve as the basis for scheduling, for communicating the construction plan and for construction monitoring. Completion of tasks will also often serve as a basis for progress payments from the owner. Thus, more detailed task definitions can be quite useful.

But more detailed task breakdowns are only valuable to the extent that the resources required, durations and activity relationships are realistically estimated for each activity. Providing detailed work task breakdowns is not helpful without a commensurate effort to provide realistic resource requirement estimates.

As more powerful, computer-based scheduling and monitoring procedures are introduced, the ease of defining and manipulating tasks will increase, and the number of work tasks can reasonably be expected to expand.

TASK DEFINITION FOR A ROAD BUILDING PROJECT

As an example of construction planning, suppose that we wish to develop a plan for a road construction project including two culverts. Initially, we divide project activities into three categories as shown in Figure 7.2: structures, roadway, and general. This division is based on the major types of design elements to be constructed. Within the roadway work, a further sub-division is into earthwork and pavement.

Within these subdivisions, we identify clearing, excavation, filling and finishing (including seeding and sodding) associated with earthwork, and we define watering, compaction and paving sub-activities associated with pavement. Finally, we note that the roadway segment is fairly long, and so individual activities can be defined for different physical segments along the roadway path. In Figure 7.2, we divide each paving and earthwork activity into activities specific to each of two roadway segments.

For the culvert construction, we define the sub-divisions of structural excavation, concreting, and reinforcing. Even more specifically, structural excavation is divided into excavation itself and the required backfill and compaction.

Similarly, concreting is divided into placing concrete forms, pouring concrete, stripping forms, and curing the concrete. As a final step in the structural planning, detailed activities are defined for reinforcing each of the two culverts. General work activities are defined for move in, general supervision, and clean up. As a result of this planning, over thirty different detailed activities have been defined.

At the option of the planner, additional activities might also be defined for this project. For example, materials ordering or lane striping might be included as separate activities. It might also be the case that a planner would define a different hierarchy of work breakdowns than that shown in Figure 7.2. For example, placing reinforcing might have been a sub-activity under concreting for culverts.

One reason for separating reinforcement placement might be to emphasize the different material and resources required for this activity. Also, the division into separate roadway segments and culverts might have been introduced early in the hierarchy. With all these potential differences, the important aspect is to insure that all necessary activities are included somewhere in the final plan.

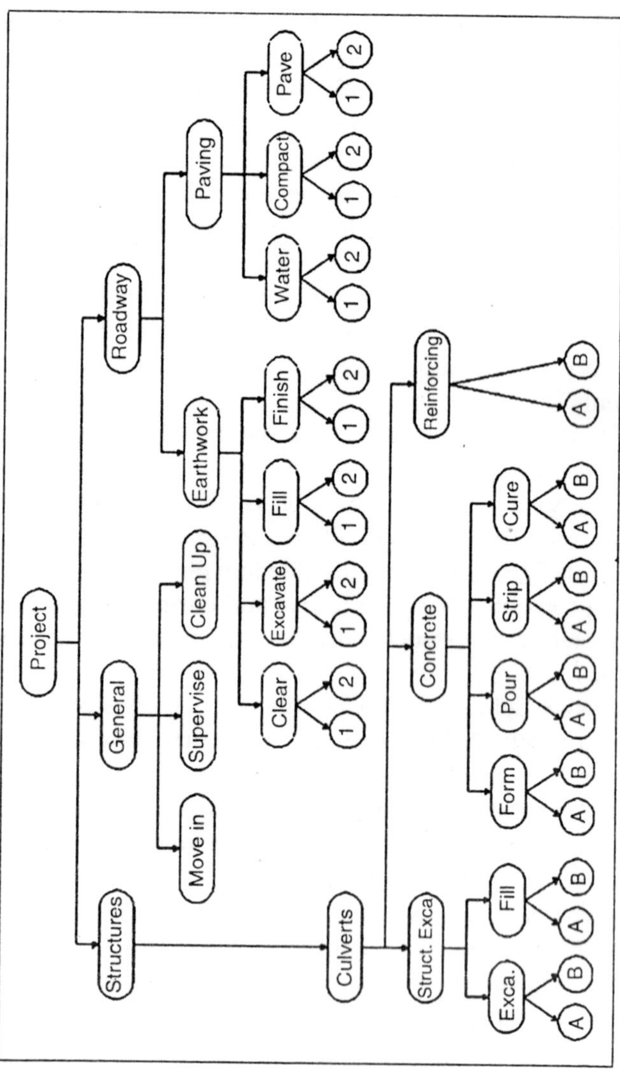

Fig. 7.2 Illustrative Hierarchical Activity Divisions for a Roadway Project

ACTIVITIES

Once work activities have been defined, the relationships

among the activities can be specified. *Precedence* relations between activities signify that the activities must take place in a particular sequence. Numerous natural sequences exist for construction activities due to requirements for structural integrity, regulations, and other technical requirements. For example, design drawings cannot be checked before they are drawn.

Diagrammatically, precedence relationships can be illustrated by a *network* or *graph* in which the activities are represented by arrows as in Figure 7.3. The arrows in Figure 7.3 are called *branches* or *links* in the *activity network*, while the circles marking the beginning or end of each arrow are called *nodes* or *events*.

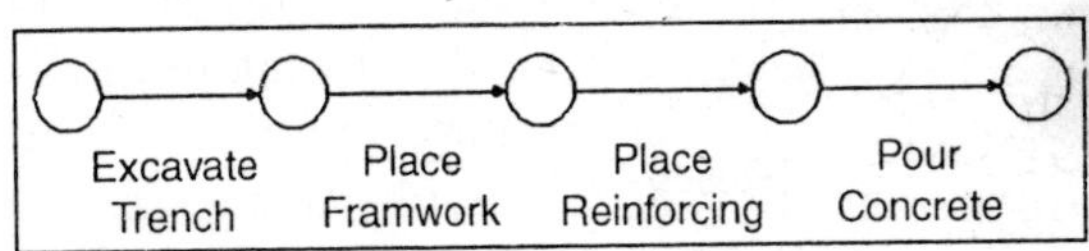

Fig. 7.3 Illustrative Set of Four Activities with Precedence's

More complicated precedence relationships can also be specified. For example, one activity might not be able to start for several days after the completion of another activity. As a common example, concrete might have to cure (or set) for several days before formwork is removed. This restriction on the removal of forms activity is called a *lag* between the completion of one activity (*i.e.*, pouring concrete in this case) and the start of another activity (*i.e.*, removing formwork in this case). Many computer based scheduling programmes permit the use of a variety of precedence relationships.

Three mistakes should be avoided in specifying predecessor relationships for construction plans. First, a circle of activity precedence's will result in an impossible plan. For example, if activity A precedes activity B, activity B precedes activity C, and activity C precedes activity A, then the project can never be started or completed! Figure 7.4 illustrates the resulting activity network. Fortunately, formal scheduling methods and good computer scheduling programmes will find any such errors in the logic of the construction plan.

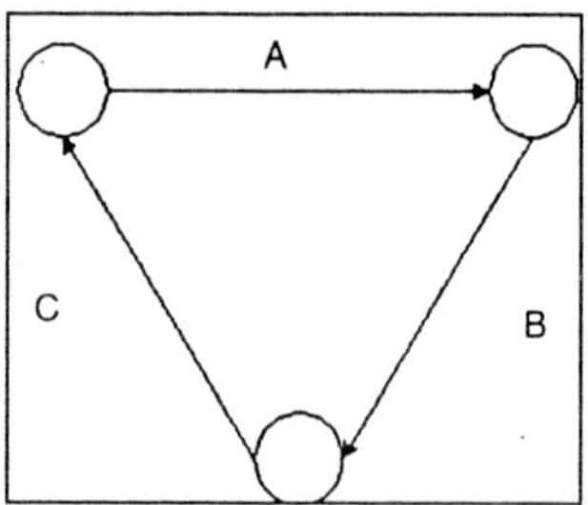

Fig. 7.4 Example of an Impossible Work Plan

Forgetting a necessary precedence relationship can be more insidious. For example, suppose that installation of dry wall should be done prior to floor finishing. Ignoring this precedence relationship may result in both activities being scheduled at the same time. Corrections on the spot may result in increased costs or problems of quality in the completed project.

Unfortunately, there are few ways in which precedence omissions can be found other than with checks by knowledgeable managers or by comparison to comparable projects. One other possible but little used mechanism for checking precedence is to conduct a physical or computer based simulation of the construction process and observe any problems.

Finally, it is important to realise that different types of precedence relationships can be defined and that each has different implications for the schedule of activities:

- Some activities have a necessary technical or physical relationship that cannot be superseded. For example, concrete pours cannot proceed before formwork and reinforcement are in place.
- Some activities have a necessary precedence relationship over a continuous space rather than as discrete work task relationships. For example, formwork may be placed in the first part of an excavation trench even as the excavation equipment

continues to work further along in the trench. Formwork placement cannot proceed further than the excavation, but the two activities can be started and stopped independently within this constraint.

- Some "precedence relationships" are not technically necessary but are imposed due to implicit decisions within the construction plan. For example, two activities may require the same piece of equipment so a precedence relationship might be defined between the two to insure that they are not scheduled for the same time period. Which activity is scheduled first is arbitrary. As a second example, reversing the sequence of two activities may be technically possible but more expensive. In this case, the precedence relationship is not physically necessary but only applied to reduce costs as perceived at the time of scheduling.

In revising schedules as work proceeds, it is important to realise that different types of precedence relationships have quite different implications for the flexibility and cost of changing the construction plan. Unfortunately, many formal scheduling systems do not possess the capability of indicating this type of flexibility. As a result, the burden is placed upon the manager of making such decisions and insuring realistic and effective schedules. With all the other responsibilities of a project manager, it is no surprise that preparing or revising the formal, computer based construction plan is a low priority to a manager in such cases. Nevertheless, formal construction plans may be essential for good management of complicated projects.

SITE PREPARATION AND FOUNDATION WORK

Suppose that a site preparation and concrete slab foundation construction project consists of nine different activities:

A. Site clearing (of brush and minor debris),
B. Removal of trees,
C. General excavation,
D. Grading general area,
E. Excavation for utility trenches,

F. Placing formwork and reinforcement for concrete,
G. Installing sewer lines,
H. Installing other utilities,
I. Pouring concrete.

Activities A (site clearing) and B (tree removal) do not have preceding activities since they depend on none of the other activities. We assume that activities C (general excavation) and D (general grading) are preceded by activity A (site clearing). It might also be the case that the planner wished to delay any excavation until trees were removed, so that B (tree removal) would be a precedent activity to C (general excavation) and D (general grading).

Activities E (trench excavation) and F (concrete preparation) cannot begin until the completion of general excavation and tree removal, since they involve subsequent excavation and trench preparation. Activities G (install lines) and H (install utilities) represent installation in the utility trenches and cannot be attempted until the trenches are prepared, so that activity E (trench excavation) is a preceding activity. We also assume that the utilities should not be installed until grading is completed to avoid equipment conflicts, so activity D (general grading) is also preceding activities G (install sewers) and H (install utilities).

Finally, activity I (pour concrete) cannot begin until the sewer line is installed and formwork and reinforcement are ready, so activities F and G are preceding. Other utilities may be routed over the slab foundation, so activity H (install utilities) is not necessarily a preceding activity for activity I (pour concrete). The result of our planning are the immediate precedences shown in Table.

Table. Precedence Relations for a Nine-Activity Project Example

Activity	Description	Predecessors
A	Site clearing	—
B	Removal of trees	—
C	General excavation	A
D	Grading general area	A

E	Excavation for utility trenches	B,C
F	Placing formwork and reinforcement for concrete	B,C
G	Installing sewer lines	D,E
H	Installing other utilities	D,E
I	Pouring concrete	F,G

With this information, the next problem is to represent the activities in a network diagram and to determine all the precedence relationships among the activities. One network representation of these nine activities is shown in Figure 7.5, in which the activities appear as branches or links between nodes. The nodes represent milestones of possible beginning and starting times. This representation is called an *activity-on-branch* diagram. Note that an initial event beginning activity is defined, while node 5 represents the completion of all activities.

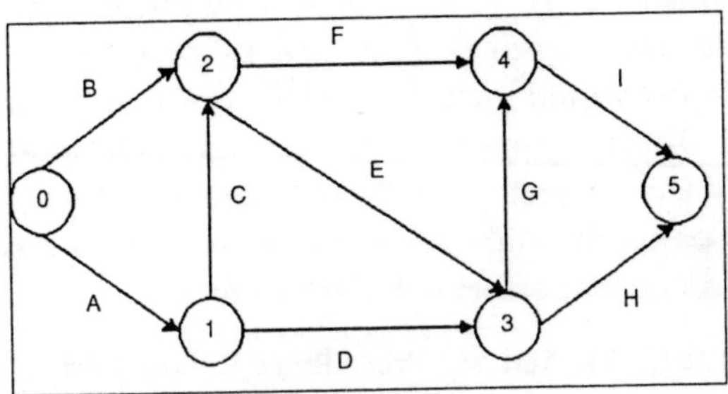

Fig. 7.5 Activity-on-Branch Representation of a Nine Activity Project

Alternatively, the nine activities could be represented by nodes and predecessor relationships by branches or links, as in Figure 7.6. The result is an*activity-on-node* diagram. In Figure 7.5, new activity nodes representing the beginning and the end of construction have been added to mark these important milestones. These network representations of activities can be very helpful in visualizing the various activities and their relationships for a project. Whether activities are represented as branches or as nodes is largely a matter of organizational or personal choice.

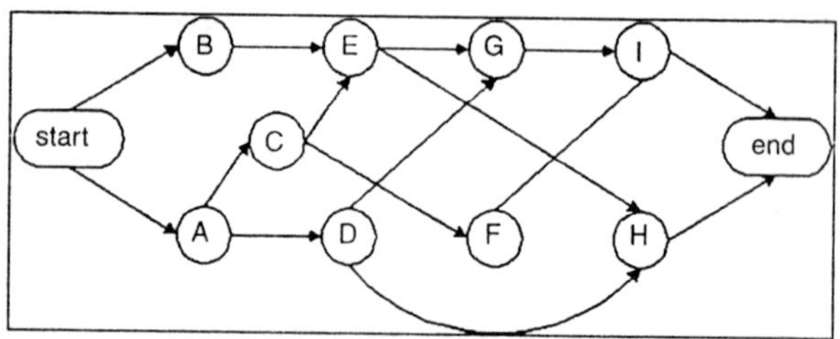

Fig. 7.6 Activity-on-Node Representation of a Nine Activity Project

It is also notable that Table lists only the *immediate* predecessor relationships. Clearly, there are other precedence relationships which involve more than one activity. For example, "installing sewer lines" (activity G) cannot be undertaken before "site clearing" (Activity A) is complete since the activity "grading general area" (Activity D) must precede activity G and must follow activity A. Table is an *implicit* precedence list since only immediate predecessors are recorded. An explicit predecessor list would include *all* of the preceding activities for activity G. Table shows all such predecessor relationships implied by the project plan. This table can be produced by tracing all paths through the network back from a particular activity and can be performed algorithmic.

Table. All Activity Precedence Relationships for a Nine-Activity Project

Predecess or Activity	Direct Successor Activities	All Successor Activities	All Predecessor Activities
A	C,D	E,F,G,H,I	---
B	E,F	G,H,I	---
C	E,F	G,H,I	A
D	G,H	I	A
E	G,H	I	A,B,C
F	I	---	A,B,C
G	I	---	A,B,C,D,E
H	---	---	A,B,C,D,E
I	---	---	A,B,C,D,E,F,G

ESTIMATING ACTIVITY DURATIONS

In most scheduling procedures, each work activity has an associated time duration. These durations are used extensively in preparing a schedule. For example, suppose that the durations shown in Table were estimated for the project. The entire set of activities would then require at least 3 days, since the activities follow one another directly and require a total of 1.0 + 0.5 + 0.5 + 1.0 = 3 days. If another activity proceeded in *parallel* with this sequence, the 3 day minimum duration of these four activities is unaffected. More than 3 days would be required for the sequence if there was a delay or a lag between the completion of one activity and the start of another.

Table. Durations and Predecessors for a Four Activity Project Illustration

Activity	Predecessor	Duration (Days)
Excavate trench	---	1.0
Place formwork	Excavate trench	0.5
Place reinforcing	Place formwork	0.5
Pour concrete	Place reinforcing	1.0

All formal scheduling procedures rely upon estimates of the durations of the various project activities as well as the definitions of the predecessor relationships among tasks. The variability of an activity's duration may also be considered. Formally, the *probability distribution* of an activity's duration as well as the expected or most likely duration may be used in scheduling. A probability distribution indicates the chance that a particular activity duration will occur. In advance of actually doing a particular task, we cannot be certain exactly how long the task will require.

A straightforward approach to the estimation of activity durations is to keep historical records of particular activities and rely on the average durations from this experience in making new duration estimates. Since the scope of activities are unlikely to be identical between different projects, unit productivity rates are typically employed for this purpose.

For example, the duration of an activity D_{ij} such as concrete formwork assembly might be estimated as:

$$D_{ij} = \frac{A_{ij}}{P_{ij}N_{ij}}$$

where A_{ij} is the required formwork area to assemble (in square yards), P_{ij} is the average productivity of a standard crew in this task (measured in square yards per hour), and N_{ij} is the number of crews assigned to the task. In some organizations, unit production time, T_{ij}, is defined as the time required to complete a unit of work by a standard crew (measured in hours per square yards) is used as a productivity measure such that T_{ij} is a reciprocal of P_{ij}.

A formula such as Eq. can be used for nearly all construction activities. Typically, the required quantity of work, A_{ij} is determined from detailed examination of the final facility design. This *quantity-take-off* to obtain the required amounts of materials, volumes, and areas is a very common process in bid preparation by contractors. In some countries, specialized quantity surveyors provide the information on required quantities for all potential contractors and the owner.

The number of crews working, N_{ij}, is decided by the planner. In many cases, the number or amount of resources applied to particular activities may be modified in light of the resulting project plan and schedule. Finally, some estimate of the expected work productivity, P_{ij} must be provided to apply Equation. As with cost factors, commercial services can provide average productivity figure 7.7 for many standard activities of this sort. Historical records in a firm can also provide data for estimation of productivities.

The calculation of a duration as in Equation is only an approximation to the actual activity duration for a number of reasons. First, it is usually the case that peculiarities of the project make the accomplishment of a particular activity more or less difficult. For example, access to the forms in a particular location may be difficult; as a result, the productivity of assembling forms may be *lower* than the average value for a particular project. Often, adjustments based on engineering judgment are made to the calculated durations from Equation for this reason.

In addition, productivity rates may vary in both systematic and random fashions from the average. An example of systematic variation is the effect of *learning* on productivity. As a crew becomes familiar with an activity and the work habits of the crew, their productivity will typically improve. Figure 7.7 illustrates the type of productivity increase that might occur with experience; this curve is called a *learning curve*.

The result is that productivity P_{ij} is a function of the duration of an activity or project. A common construction example is that the assembly of floors in a building might go faster at higher levels due to improved productivity even though the transportation time up to the active construction area is longer. Again, historical records or subjective adjustments might be made to represent learning curve variations in average productivity.

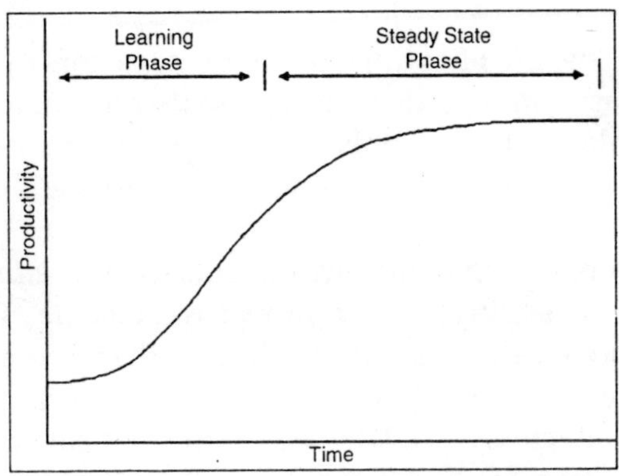

Fig. 7.7 Illustration of Productivity Changes Due to Learning

Random factors will also influence productivity rates and make estimation of activity durations uncertain. For example, a scheduler will typically not know at the time of making the initial schedule how skillful the crew and manager will be that are assigned to a particular project. The productivity of a skilled designer may be many times that of an unskilled engineer. In the absence of specific knowledge, the estimator can only use average values of productivity.

Weather effects are often very important and thus deserve particular attention in estimating durations. Weather has both systematic and random influences on activity durations. Whether or not a rainstorm will come on a particular day is certainly a random effect that will influence the productivity of many activities. However, the likelihood of a rainstorm is likely to vary systematically from one month or one site to the next.

Adjustment factors for inclement weather as well as meteorological records can be used to incorporate the effects of weather on durations. As a simple example, an activity might require ten days in perfect weather, but the activity could not proceed in the rain. Furthermore, suppose that rain is expected ten per cent of the days in a particular month. In this case, the expected activity duration is eleven days including one expected rain day.

Finally, the use of average productivity factors themselves cause problems in the calculation presented in Equation. The expected value of the multiplicative reciprocal of a variable is not exactly equal to the reciprocal of the variable's expected value.

For example, if productivity on an activity is either six in good weather (ie., $P=6$) or two in bad weather (ie., $P=2$) and good or bad weather is equally likely, then the expected productivity is $P = (6)(0.5) + (2)(0.5) = 4$, and the reciprocal of expected productivity is 1/4. However, the expected reciprocal of productivity is $E[1/P] = (0.5)/6 + (0.5)/2 = 1/3$.

The reciprocal of expected productivity is 25% less than the expected value of the reciprocal in this case! By representing only two possible productivity values, this example represents an extreme case, but it is always true that the use of average productivity factors in Equation will result in *optimistic* estimates of activity durations. The use of actual averages for the reciprocals of productivity or small adjustment factors may be used to correct for this non-linearity problem. The simple duration calculation shown in Equation also assumes an inverse linear relationship between the number of crews assigned to an activity and the total duration of work. While this is a reasonable assumption in situations for which crews can work independently and require no special coordination, it need not always be true. For example, design tasks may be divided among numerous architects and engineers, but delays to insure proper coordination and communication increase as the number of workers increase.

As another example, insuring a smooth flow of material to all crews on a site may be increasingly difficult as the number of crews increase. In these latter cases, the relationship between activity duration and the number of crews is unlikely to be inversely proportional as shown in Equation. As a result, adjustments to the estimated productivity from Equation must be made.

Alternatively, more complicated functional relationships might be estimated between duration and resources used in the same way that nonlinear preliminary or conceptual cost estimate models are prepared.

One mechanism to formalize the estimation of activity durations is to employ a hierarchical estimation framework. This approach decomposes the estimation problem into component parts in which the higher levels in the hierarchy represent attributes which depend upon the details of lower level adjustments and calculations.

For example, Figure 7.8 represents various levels in the estimation of the duration of masonry construction. At the lowest level, the maximum productivity for the activity is estimated based upon general work conditions.

Table illustrates some possible maximum productivity values that might be employed in this estimation. At the next higher level, adjustments to these maximum productivities are made to account for special site conditions and crew compositions; table illustrates some possible adjustment rules.

At the highest level, adjustments for overall effects such as weather are introduced. Also shown in Figure 7.8 are nodes to estimate down or unproductive time associated with the masonry construction activity.

The formalization of the estimation process illustrated in Figure 7.8 permits the development of computer aids for the estimation process or can serve as a conceptual framework for a human estimator.

Tablr. Maximum Productivity Estimates for Masonry Work

Masonry unit size	**Condition(s)**	**Maximum produstivity achievable-**
8 inch block	None	400units/ day/mason
6 inch	Wall is "long"	430units/ day/mason
6 inch	Wall is not "long"	370 units/ day/mason
12 inch	Labor is nonunion	300 units/ day/mason
4 inch	Wall is "long" Weather is "warm and dry" or high-strength mortar is used	480 units/ day/mason
4 inch	Wall is not "long" Weather is "warm and dry" or high-strength mortar is used	430 units/ day/mason
4 inch	Wall is "long" Weather is not "warm and dry" or high-strength mortar is not used	370 units/ day/mason
4 inch	Wall is not "long" Weather is not "warm and dry" or high-strength mortar is not used	320 units/ day/mason
8 inch	There is support from existing wall	1,000units/ day/mason

8 inch	There is no support from existing wall	750 units/ day/mason
12 inch	There is support from existing wall	700 units/ day/mason
12 inch	There is no support from existing wall	550

Table. Possible Adjustments to Maximum Productivities for Masonry Construction/caption>

Impact	**Condition(s)**	**Adjustment magnitude (% of maximum)**
Crew type	Crew type is nonunion Job is "large"	15%
Crew type	Crew type is union Job is "small"	10%
Supporting laboUr	There are less than two laborers per crew	20%
Supporting laboUr	There are more than two masons/laborers	10%
Elevation	Steel frame building with masonry exterior wall has "insufficient" support labor	10%
Elevation	Solid masonry building with work on exterior uses nonunion laboUr	12%
Visibility	Block is not covered	7%
Temperature	Temperature is below 45° F	15%
Temperature	Temperature is above 45° F	10%
Brick texture	Bricks are baked high Weather is cold or moist	10%

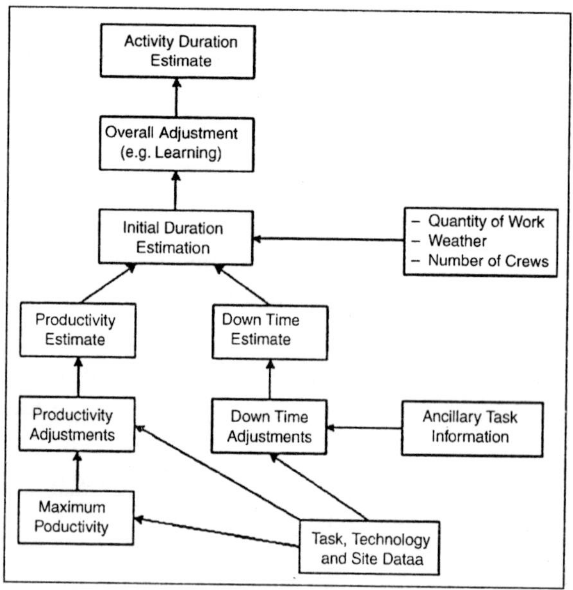

Fig. 7.8 A Hierarchical Estimation Framework for Masonry Construction

In addition to the problem of estimating the expected duration of an activity, some scheduling procedures explicitly consider the uncertainty in activity duration estimates by using the probabilistic distribution of activity durations. That is, the duration of a particular activity is assu med to be a random variable that is distributed in a particular fashion.For example, an activity duration might be assumed to be distributed as a normal or a beta distributed random variable as illustrated in Figure 7.9. This figure 7.9 shows the probability or chance of experiencing a particular activity duration based on a probabilistic distribution.

The beta distribution is often used to characterize activity durations, since it can have an absolute minimum and an absolute maximum of possible duration times. The normal distribution is a good approximation to the beta distribution in the center of the distribution and is easy to work with, so it is often used as an approximation.

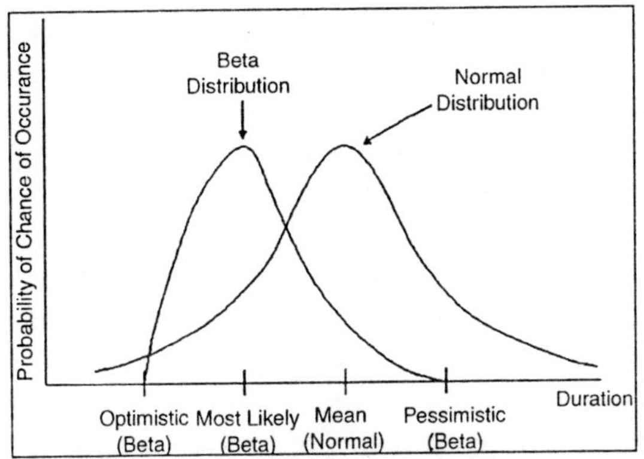

Fig. 7.9 Beta and Normally Distributed Activity Durations

If a standard random variable is used to characterize the distribution of activity durations, then only a few parameters are required to calculate the probability of any particular duration.

Still, the estimation problem is increased considerably since more than one parameter is required to characterize most of the probabilistic distribution used to represent activity durations. For the beta distribution, three or four parameters are required depending on its generality, whereas the normal distribution requires two parameters.

As an example, the normal distribution is characterized by two parameters, μ and σ representing the average duration and the standard deviation of the duration, respectively. Alternatively, the *variance* of the distribution σ^2 could be used to describe or characterize the variability of duration times; the variance is the value of the standard deviation multiplied by itself. From historical data, these two parameters can be estimated as:

$$\mu \approx \overline{x} = \sum_{k=1}^{n} \frac{x_k}{n}$$

$$\sigma^2 \approx \sum_{k=1}^{n} \frac{(x_k - \overline{x})^2}{n-1}$$

where we assume that n different observations x_k of the random variable x are available. This estimation process might be applied to activity durations directly (so that x_k would be a record of an activity duration D_{ij} on a past project) or to the estimation of the distribution of productivities (so that x_k would be a record of the productivity in an activity P_i) on a past project) which, in turn, is used to estimate durations using Equation. If more accuracy is desired, the estimation equations for mean and standard deviation, Equations would be used to estimate the mean and standard deviation of the reciprocal of productivity to avoid non-linear effects. Using estimates of productivities, the standard deviation of activity duration would be calculated as:

$$\sigma_{ij} \approx \frac{A_{ij}\sigma_{1/p}}{N_{ij}}$$

where $\sigma_{1/p}$ is the estimated standard deviation of the reciprocal of productivity that is calculated from Equation by substituting 1/P for x.

ESTIMATING RESOURCE REQUIREMENTS FOR WORK ACTIVITIES

In addition to precedence relationships and time durations, *resource requirements* are usually estimated for each activity. Since the work activities defined for a project are comprehensive, the total resources required for the project are the sum of the resources required for the various activities. By making resource requirement estimates for each activity, the requirements for particular resources during the course of the project can be identified. Potential bottlenecks can thus be identified, and schedule, resource allocation or technology changes made to avoid problems.

Many formal scheduling procedures can incorporate constraints imposed by the availability of particular resources. For example, the unavailability of a specific piece of equipment or crew may prohibit activities from being undertaken at a particular time. Another type of resource is space. A planner typically will schedule only one activity in the same location at the same time. While activities requiring the same space may have no necessary technical precedence, simultaneous work might not be possible.

The initial problem in estimating resource requirements is to decide the extent and number of resources that might be defined. At a very aggregate level, resources categories might be limited to the amount of labor (measured in man-hours or in dollars), the amount of materials required for an activity, and the total cost of the activity. At this aggregate level, the resource estimates may be useful for purposes of project monitoring and cash flow planning. For example, actual expenditures on an activity can be compared with the estimated required resources to reveal any problems that are being encountered during the course of a project. However, this aggregate definition of resource use would not reveal bottlenecks associated with particular types of equipment or workers.

More detailed definitions of required resources would include the number and type of both workers and equipment required by an activity as well as the amount and types of materials. Standard resource requirements for particular activities can be recorded and adjusted for the special conditions of particular projects. As a result, the resources types required for particular activities may already be defined. Reliance on historical or standard activity definitions of this type requires a standard coding system for activities.

In making adjustments for the resources required by a particular activity, most of the problems encountered in forming duration estimations described in the previous section are also present. In particular, resources such as labor requirements will vary in proportion to the work productivity, P_{ij}, used to estimate activity durations in Equation.

Mathematically, a typical estimating equation would be:

$$R_{ij}^{k} = D_{ij} N_{ij} U_{ij}^{k}$$

where R^k_{ij} are the resources of type k required by activity ij, D_{ij} is the duration of activity ij, N_{ij} is the number of standard crews allocated to activity ij, and U^k_{ij} is the amount of resource type k used per standard crew. For example, if an activity required eight hours with two crews assigned and each crew required three workers, the effort would be R = 8*2*3 = 48 labor-hours.

From the planning perspective, the important decisions in estimating resource requirements are to determine the type of technology and equipment to employ and the number of crews to allocate to each task. Clearly, assigning additional crews might result in faster completion of a particular activity. However, additional crews might result in congestion and coördination problems, so that work productivity might decline.

Resource Requirements for Block Foundations

In placing concrete block foundation walls, a typical crew would consist of three bricklayers and two bricklayer helpers. If sufficient space was available on the site, several crews could work on the same job at the same time, thereby speeding up completion of the activity in proportion to the number of crews.

In more restricted sites, multiple crews might interfere with one another. For special considerations such as complicated scaffolding or large blocks (such as twelve inch block), a bricklayer helper for each bricklayer might be required to insure smooth and productive work. In general, standard crew composition depends upon the specific construction task and the equipment or technology employed. These standard crews are then adjusted in response to special characteristics of a particular site.

Pouring Concrete Slabs

For large concrete pours on horizontal slabs, it is important to plan the activity so that the slab for a full block can be completed continuously in a single day. Resources required for pouring the concrete depend upon the technology used. For example, a standard crew for pumping concrete to the slab might include a foreman, five laborers, one finisher, and one equipment operator. Related equipment would be vibrators and the concrete pump itself. For delivering concrete with a chute directly from the delivery truck, the standard crew might consist of a foreman, four laborers and a finisher. The number of crews would be chosen to insure that the desired amount of concrete could be placed in a single day. In addition to the resources involved in the actual placement, it would also be necessary to insure a sufficient number of delivery trucks and availability of the concrete itself.

CODING SYSTEMS

One objective in many construction planning efforts is to define the plan within the constraints of a universal *coding system* for identifying activities. Each activity defined for a project would be identified by a pre-defined code specific to that activity. The use of a common nomenclature or identification system is basically motivated by the desire for better integration of organizational efforts and improved information flow. In particular, coding systems are adopted to provide a numbering system to replace verbal descriptions of items.

These codes reduce the length or complexity of the information to be recorded. A common coding system within an organization also aids consistency in definitions and categories between projects and among the various parties involved in a project. Common coding systems also aid in the retrieval of historical records of cost, productivity and duration on particular activities.

In North America, the most widely used standard coding system for constructed facilities is the MASTERFORMAT system developed by the Construction Specifications Institute (CSI) of the United States and Construction Specifications of Canada. After development of separate systems, this combined system was originally introduced as the Uniform Construction Index (UCI) in 1972 and was subsequently adopted for use by numerous firms, information providers, professional societies and trade organizations. The term MASTERFORMAT was introduced with the 1978 revision of the UCI codes. MASTERFORMAT provides a standard identification code for nearly all the elements associated with building construction. MASTERFORMAT involves a hierarchical coding system with multiple levels plus keyword text descriptions of each item. In the numerical coding system, the first two digits represent one of the sixteen divisions for work; a seventeenth division is used to code conditions of the contract for a constructor. In the latest version of the MASTERFORMAT, a third digit is added to indicate a subdivision within each division. Each division is further specified by a three digit extension indicating another level of subdivisions. In many cases, these subdivisions are further divided with an additional three digits to identify more specific work items or materials.

For example, the code 16-950-960, "Electrical Equipment Testing" are defined as within Division 16 (Electrical) and Sub-Division 950 (Testing). The keywords "Electrical Equipment Testing" is a standard description of the activity. The seventeen major divisions in the UCI/CSI MASTERFORMAT system are shown in Table. As an example, site work second level divisions are shown in Table.

Major divisions in the uniform construction index:

- Conditions of the contract
- General requirements
- Site work
- Concrete
- Masonry
- Metals
- Wood and plastics
- Thermal and moisture prevention
- Doors and windows
- Finishes
- Specialties
- Equipment
- Furnishings
- Special construction
- Conveying system
- Mechanical
- Electrical

While MASTERFORMAT provides a very useful means of organizing and communicating information, it has some obvious limitations as a complete project coding system. First, more specific information such as location of work or responsible organization might be required for project cost control. Code extensions are then added in addition to the digits in the basic MASTERFORMAT codes.

For example, a typical extended code might have the following elements:

0534.02220.21.A.00.cf34

The first four digits indicate the project for this activity; this code refers to an activity on project number 0534. The next five digits refer to the MASTERFORMAT secondary division; referring to Table, this activity would be 02220 "Excavating, Backfilling and Compacting." The next two digits refer to specific activities defined within this MASTERFORMAT code; the digits 21 in this example might refer to excavation of column footings. The next character refers to the *block* or general area on the site that the activity will take place; in this case, block A is indicated. The digits 00 could be replaced by a code to indicate the responsible organization for the activity. Finally, the characters cf34 refer to the particular design element number for which this excavation is intended; in this case, column footing number 34 is intended.

Thus, this activity is to perform the excavation for column footing number 34 in block A on the site. Note that a number of additional activities would be associated with column footing 34, including formwork and concreting. Additional fields in the coding systems might also be added to indicate the responsible crew for this activity or to identify the specific location of the activity on the site (defined, for example, as x, y and z coordinates with respect to a base point).

As a second problem, the MASTERFORMAT system was originally designed for building construction activities, so it is difficult to include various construction activities for other types of facilities or activities associated with planning or design. Different coding systems have been provided by other organizations in particular sub-fields such as power plants or roadways. Nevertheless, MASTERFORMAT provides a useful starting point for organizing information in different construction domains.

In devising organizational codes for project activities, there is a continual tension between adopting systems that are convenient or expedient for one project or for one project manager and systems appropriate for an entire organization. As a general rule, the record keeping and communication advantages of standard systems are excellent arguments for their adoption. Even in small projects, however, ad hoc or haphazard coding systems can lead to problems as the system is revised and extended over time.

Table. Secondary Divisions in MASTERFORMAT for Site Work

- 02-010: Subsurface investigation
- 02-012: Standard penetration tests
- 02-016: Seismic investigation
- 02-050: Demolition
- 02-060: Building demolition
- 02-070: Selective demolition
- 02-075: Concrete removal
- 02-080: Concrete removal
- 02-100: Site preparation
- 02-110: Site clearing
- 02-115: Selective clearing
- 02-120: Structure moving
- 02-140: Dewatering
- 02-150: Shoring and underpinning
- 02-160: Excavation supporting system
- 02-170: Cofferdams
- 02-200: Earthwork
- 02-210: Grading
- 02-220: Excavating, backfilling and compaction
- 02-230: Base course
- 02-240: Soil stabilization
- 02-250: Vibro-floatation
- 02-270: Slope protection
- 02-280: Soil treatment

- 02-290: Earth dams
- 02-300: Tunneling
- 02-305: Tunnel ventilation
- 02-310: Tunnel excavating
- 02-320: Tunnel lining
- 02-330: Tunnel grouting
- 02-340: Tunnel support systems
- 02-350: Piles and caissons
- 02-355: Pile driving
- 02-360: Driven piles
- 02-370: Bored/augered piles
- 02-380: Caissons
- 02-450: Railroad work
- 02-480: Marine work
- 02-500: Paving and surfacing
- 02-510: Walk, road and parking paving
- 02-515: Unit pavers
- 02-525: Curbs
- 02-530: Athletic paving and surfacing
- 02-540: Synthetic surfacing
- 02-545: Surfacing
- 02-550: Highway paving
- 02-560: Airfield paving
- 02-575: Pavement repair
- 02-580: Pavement marking
- 02-600: Piped utility materials
- 02-660: Water distribution
- 02-680: Fuel distribution
- 02-700: Sewage and drainage
- 02-760: Restoration of underground pipelines
- 02-770: Ponds and reservoirs
- 02-800: Power and communications
- 02-880: Site improvements
- 02-900: Landscaping

8

Quality Control and Safety

QUALITY AND SAFETY CONCERNS IN CONSTRUCTION

Quality control and safety represent increasingly important concerns for project managers. Defects or failures in constructed facilities can result in very large costs. Even with minor defects, re-construction may be required and facility operations impaired. Increased costs and delays are the result. In the worst case, failures may cause personal injuries or fatalities. Accidents during the construction process can similarly result in personal injuries and large costs. Indirect costs of insurance, inspection and regulation are increasing rapidly due to these increased direct costs. Good project managers try to ensure that the job is done right the first time and that no major accidents occur on the project.

As with cost control, the most important decisions regarding the quality of a completed facility are made during the design and planning stages rather than during construction. It is during these preliminary stages that component configurations, material specifications and functional performance are decided. Quality control during construction consists largely of insuring *conformance* to these original design and planning decisions. While conformance to existing design decisions is the primary focus of quality control, there are exceptions to this rule. First, unforeseen circumstances, incorrect design decisions or changes desired by an owner in the facility function may require re-evaluation of design decisions during the course of construction. While these changes may be motivated by the concern for quality, they represent occasions for re-design with all the attendant objectives and constraints. As a second case, some designs rely upon informed and appropriate decision making during the construction process itself. For example, some tunneling methods make decisions about the amount of shoring required at different locations based upon observation of soil conditions during the tunneling process. Since such decisions are based on better information concerning actual site conditions, the facility design may be more cost effective as a result.

With the attention to conformance as the measure of quality during the construction process, the specification of quality requirements in the design and contract documentation becomes extremely important. Quality requirements should be clear and verifiable, so that all parties in the project can understand the requirements for conformance. Much of the discussion in this chapter relates to the development and the implications of different quality requirements for construction as well as the issues associated with insuring conformance.

Safety during the construction project is also influenced in large part by decisions made during the planning and design process. Some designs or construction plans are inherently difficult and dangerous to implement, whereas other, comparable plans may considerably reduce the possibility of accidents. For example, clear separation of traffic from construction zones during roadway rehabilitation can greatly reduce the possibility of accidental collisions. Beyond these design decisions, safety largely depends upon education, vigilance and cooperation during the construction process. Workers should be constantly alert to the possibilities of accidents and avoid taken unnecessary risks.

ORGANIZING FOR QUALITY AND SAFETY

A variety of different organizations are possible for quality and safety control during construction. One common model is to have a group responsible for quality assurance and another group primarily responsible for safety within an organization. In large organizations, departments dedicated to quality assurance and to safety might assign specific individuals to assume responsibility for these functions on particular projects. For smaller projects, the project manager or an assistant might assume these and other responsibilities. In either case, insuring safe and quality construction is a concern of the project manager in overall charge of the project in addition to the concerns of personnel, cost, time and other management issues.

Inspectors and quality assurance personnel will be involved in a project to represent a variety of different organizations. Each of the parties directly concerned with the project may have their own quality and safety inspectors, including the owner, the engineer/architect, and the various constructor firms. These inspectors may be contractors from specialized quality assurance organizations. In addition to on-site inspections, samples of materials will commonly be tested by specialized laboratories to insure compliance. Inspectors to insure compliance with regulatory requirements will also be involved. Common examples are inspectors for the local government's building department, for environmental agencies, and for occupational health and safety agencies.

The US Occupational Safety and Health Administration (OSHA) routinely conducts site visits of work places in conjunction with approved state inspection agencies. OSHA inspectors are required by law to issue citations for all standard violations observed. Safety standards prescribe a variety of mechanical safeguards and procedures; for example, ladder safety is covered by over 140 regulations. In cases of extreme non-compliance with standards, OSHA inspectors can stop work on a project. However, only a small fraction of construction sites are visited by OSHA inspectors and most construction site accidents are not caused by violations of existing standards. As a result, safety is largely the responsibility of the managers on site rather than that of public inspectors.

While the multitude of participants involved in the construction process require the services of inspectors, it cannot be emphasized too strongly that inspectors are only a formal check on quality control. Quality control should be a primary objective for all the members of a project team. Managers should take responsibility for maintaining and improving quality control. Employee participation in quality control should be sought and rewarded, including the introduction of new ideas. Most important of all, quality improvement can serve as a catalyst for improved productivity. By suggesting new work methods, by avoiding rework, and by avoiding long term problems, good quality control can pay for itself. Owners should promote good quality control and seek out contractors who maintain such standards.

In addition to the various organizational bodies involved in quality control, issues of quality control arise in virtually all the functional areas of construction activities. For example, insuring accurate and useful information is an important part of maintaining quality performance. Other aspects of quality control include document control (including changes during the construction process), procurement, field inspection and testing, and final checkout of the facility.

WORK AND MATERIAL SPECIFICATIONS

Specifications of work quality are an important feature of facility designs. Specifications of required quality and components represent part of the necessary documentation to describe a facility. Typically, this documentation includes any special provisions of the facility design as well as references to generally accepted specifications to be used during construction.

General specifications of work quality are available in numerous fields and are issued in publications of organizations such as the American Society for Testing and Materials (ASTM), the American National Standards Institute (ANSI), or the Construction Specifications Institute (CSI). Distinct specifications are formalized for particular types of construction activities, such as welding standards issued by the American Welding Society, or for particular facility types, such as the *Standard Specifications for Highway Bridges* issued by the American Association of State Highway and Transportation Officials. These general specifications must be modified to reflect local conditions, policies, available materials, local regulations and other special circumstances.

Construction specifications normally consist of a series of instructions or prohibitions for specific operations. For example, the following passage illustrates a typical specification, in this case for excavation for structures:

Conform to elevations and dimensions shown on plan within a tolerance of plus or minus 0.10 foot, and extending a sufficient distance from footings and foundations to permit placing and removal of concrete formwork, installation of services, other construction, and for inspection. In excavating for footings and foundations, take care not to disturb bottom of excavation. Excavate by hand to final grade just before concrete reinforcement is placed. Trim bottoms to required lines and grades to leave solid base to receive concrete.

This set of specifications requires judgment in application since some items are not precisely specified. For example, excavation must extend a "sufficient" distance to permit inspection and other activities.

Obviously, the term "sufficient" in this case may be subject to varying interpretations. In contrast, a specification that tolerances are within plus or minus a tenth of a foot is subject to direct measurement. However, specific requirements of the facility or characteristics of the site may make the standard tolerance of a tenth of a foot inappropriate. Writing specifications typically requires a trade-off between assuming reasonable behaviour on the part of all the parties concerned in interpreting words such as "sufficient" versus the effort and possible inaccuracy in pre-specifying all operations.

In recent years, *performance specifications* have been developed for many construction operations. Rather than specifying the required construction *process*, these specifications refer to the required performance or quality of the finished facility. The exact method by which this performance is obtained is left to the construction contractor. For example, traditional specifications for asphalt pavement specified the composition of the asphalt material, the asphalt temperature during paving, and compacting procedures. In contrast, a performance specification for asphalt would detail the desired performance of the pavement with respect to impermeability, strength, etc. How the desired performance level was attained would be up to the paving contractor. In some cases, the payment for asphalt paving might increase with better quality of asphalt beyond some minimum level of performance.

CONCRETE PAVEMENT STRENGTH

Concrete pavements of superior strength result in cost savings by delaying the time at which repairs or re-construction is required. In contrast, concrete of lower quality will necessitate more frequent overlays or other repair procedures. Contract provisions with adjustments to the amount of a contractor's compensation based on pavement quality have become increasingly common in recognition of the cost savings associated with higher quality construction.

Even if a pavement does not meet the "ultimate" design standard, it is still worth using the lower quality pavement and re-surfacing later rather than completely rejecting the pavement. Based on these life cycle cost considerations, a typical pay schedule might be:

Load Ratio	Pay Factor
<0.50	Reject
0.50-0.69	0.90
0.70-0.89	0.95
0.90-1.09	1.00
1.10-1.29	1.05
1.30-1.49	1.10
>1.50	1.12

In this table, the Load Ratio is the ratio of the actual pavement strength to the desired design strength and the Pay Factor is a fraction by which the total pavement contract amount is multiplied to obtain the appropriate compensation to the contractor. For example, if a contractor achieves concrete strength twenty percent greater than the design specification, then the load ratio is 1.20 and the appropriate pay factor is 1.05, so the contractor receives a five per cent bonus. Load factors are computed after tests on the concrete actually used in a pavement.

Note that a 90% pay factor exists in this case with even pavement quality only 50% of that originally desired. This high pay factor even with weak concrete strength might exist since much of the cost of pavements are incurred in preparing the pavement foundation. Concrete strengths of less then 50% are cause for complete rejection in this case, however.

TOTAL QUALITY CONTROL

Quality control in construction typically involves insuring compliance with minimum standards of material and workmanship in order to insure the performance of the facility according to the design. These minimum standards are contained in the specifications described in the previous section. For the purpose of insuring compliance, random samples and statistical methods are commonly used as the basis for accepting or rejecting work completed and batches of materials. Rejection of a batch is based on non-conformance or violation of the relevant design specifications. Procedures for this quality control practice are described in the following sections.

An implicit assumption in these traditional quality control practices is the notion of an *acceptable quality level* which is a allowable fraction of defective items. Materials obtained from suppliers or work performed by an organization is inspected and passed as acceptable if the estimated defective percentage is within the acceptable quality level. Problems with materials or goods are corrected after delivery of the product. In contrast to this traditional approach of quality control is the goal of *total quality control*. In this system, no defective items are allowed anywhere in the construction process. While the zero defects goal can never be permanently obtained, it provides a goal so that an organization is never satisfied with its quality control programme even if defects are reduced by substantial amounts year after year.

This concept and approach to quality control was first developed in manufacturing firms in Japan and Europe, but has since spread to many construction companies. The best known formal certification for quality improvement is the International Organization for Standardization's ISO 9000 standard. ISO 9000 emphasizes good documentation, quality goals and a series of cycles of planning, implementation and review.

Total quality control is a commitment to quality expressed in all parts of an organization and typically involves many elements. Design reviews to insure safe and effective construction procedures are a major element. Other elements include extensive training for personnel, shifting the responsibility for detecting defects from quality control inspectors to workers, and continually maintaining equipment.

Worker involvement in improved quality control is often formalized in *quality circles* in which groups of workers meet regularly to make suggestions for quality improvement. Material suppliers are also required to insure zero defects in delivered goods. Initally, all materials from a supplier are inspected and batches of goods with any defective items are returned. Suppliers with good records can be certified and not subject to complete inspection subsequently.

The traditional microeconomic view of quality control is that there is an "optimum" proportion of defective items. Trying to achieve greater quality than this optimum would substantially increase costs of inspection and reduce worker productivity. However, many companies have found that commitment to total quality control has substantial economic benefits that had been unappreciated in traditional approaches. Expenses associated with inventory, rework, scrap and warranties were reduced. Worker enthusiasm and commitment improved. Customers often appreciated higher quality work and would pay a premium for good quality. As a result, improved quality control became a competitive advantage.

Of course, total quality control is difficult to apply, particular in construction. The unique nature of each facility, the variability in the workforce, the multitude of subcontractors and the cost of making necessary investments in education and procedures make programmes of total quality control in construction difficult. Nevertheless, a commitment to improved quality even without endorsing the goal of zero defects can pay real dividends to organizations.

EXPERIENCE WITH QUALITY CIRCLES

Quality circles represent a group of five to fifteen workers who meet on a frequent basis to identify, discuss and solve productivity and quality problems. A circle leader acts as liason between the workers in the group and upper levels of management.

Appearing below are some examples of reported quality circle accomplishments in construction:

- On a highway project under construction by Taisei Corporation, it was found that the loss rate of ready-mixed concrete was too high. A quality circle composed of cement masons found out that the most important reason for this was due to an inaccurate checking method. By applying the circle's recommendations, the loss rate was reduced by 11.4%.
- In a building project by Shimizu Construction Company, may cases of faulty reinforced concrete work were reported. The iron workers quality circle examined their work thoroughly and soon the faulty workmanship disappeared. A 10% increase in productivity was also achieved.

QUALITY CONTROL BY STATISTICAL METHODS

An ideal quality control programme might test all materials

and work on a particular facility. For example, non-destructive techniques such as x-ray inspection of welds can be used throughout a facility. An on-site inspector can witness the appropriateness and adequacy of construction methods at all times. Even better, individual craftsmen can perform continuing inspection of materials and their own work.

Exhaustive or 100% testing of all materials and work by inspectors can be exceedingly expensive, however. In many instances, testing requires the destruction of a material sample, so exhaustive testing is not even possible.

As a result, small samples are used to establish the basis of accepting or rejecting a particular work item or shipment of materials. Statistical methods are used to interpret the results of test on a small sample to reach a conclusion concerning the acceptability of an entire *lot* or batch of materials or work products.

The use of statistics is essential in interpreting the results of testing on a small sample. Without adequate interpretation, small sample testing results can be quite misleading. As an example, suppose that there are ten defective pieces of material in a lot of one hundred. In taking a sample of five pieces, the inspector might not find *any* defective pieces or might have *all* sample pieces defective.

Drawing a direct inference that none or all pieces in the population are defective on the basis of these samples would be incorrect. Due to this random nature of the sample selection process, testing results can vary substantially. It is only with statistical methods that issues such as the chance of different levels of defective items in the full lot can be fully analysed from a small sample test.

There are two types of statistical sampling which are commonly used for the purpose of quality control in batches of work or materials:

- The acceptance or rejection of a lot is based on the number of defective (bad) or nondefective (good) items in the sample. This is referred to as *sampling by attributes*.
- Instead of using defective and nondefective

classifications for an item, a quantitative quality measure or the value of a measured variable is used as a quality indicator. This testing procedure is referred to as *sampling by variables.*

Whatever sampling plan is used in testing, it is always assumed that the samples are representative of the entire population under consideration.

Samples are expected to be chosen randomly so that each member of the population is equally likely to be chosen. Convenient sampling plans such as sampling every twentieth piece, choosing a sample every two hours, or picking the top piece on a delivery truck may be adequate to insure a random sample if pieces are randomly mixed in a stack or in use.

However, some convenient sampling plans can be inappropriate. For example, checking only easily accessible joints in a building component is inappropriate since joints that are hard to reach may be more likely to have erection or fabrication problems.

Another assumption implicit in statistical quality control procedures is that the quality of materials or work is expected to vary from one piece to another. This is certainly true in the field of construction. While a designer may assume that all concrete is exactly the same in a building, the variations in material properties, manufacturing, handling, pouring, and temperature during setting insure that concrete is actually heterogeneous in quality.

Reducing such variations to a minimum is one aspect of quality construction. Insuring that the materials actually placed achieve some minimum quality level with respect to average properties or fraction of defectives is the task of quality control.

STATISTICAL QUALITY CONTROL WITH SAMPLING BY ATTRIBUTES

Sampling by attributes is a widely applied quality control method. The procedure is intended to determine whether or not a particular group of materials or work products is acceptable.

In the literature of statistical quality control, a group of materials or work items to be tested is called a *lot* or *batch*. An assumption in the procedure is that each item in a batch can be tested and classified as either acceptable or deficient based upon mutually acceptable testing procedures and acceptance criteria. Each lot is tested to determine if it satisfies a minimum acceptable quality level (AQL) expressed as the maximum percentage of defective items in a lot or process.

In its basic form, sampling by attributes is applied by testing a pre-defined number of sample items from a lot. If the number of defective items is greater than a trigger level, then the lot is rejected as being likely to be of unacceptable quality. Otherwise, the lot is accepted. Developing this type of *sampling plan* requires consideration of probability, statistics and acceptable risk levels on the part of the supplier and consumer of the lot.

Refinements to this basic application procedure are also possible. For example, if the number of defectives is greater than some pre-defined number, then additional sampling may be started rather than immediate rejection of the lot. In many cases, the trigger level is a single defective item in the sample. In the remainder of this section, the mathematical basis for interpreting this type of sampling plan is developed.

More formally, a lot is defined as acceptable if it contains a fraction p_1 or less defective items. Similarly, a lot is defined as unacceptable if it contains a fraction p_2 or more defective units. Generally, the acceptance fraction is less than or equal to the rejection fraction, $p_1 \leq p_2$, and the two fractions are often equal so that there is no ambiguous range of lot acceptability between p_1 and p_2.

Given a sample size and a trigger level for lot rejection or acceptance, we would like to determine the probabilities that acceptable lots might be incorrectly rejected (termed *producer's risk*) or that deficient lots might be incorrectly accepted.

Consider a lot of finite number N, in which m items are defective (bad) and the remaining (N-m) items are non-defective (good). If a random sample of n items is taken from this lot, then we can determine the probability of having different numbers of defective items in the sample. With a pre-defined acceptable number of defective items, we can then develop the probability of accepting a lot as a function of the sample size, the allowable number of defective items, and the actual fraction of defective items..

The number of different samples of size n that can be selected from a finite population N is termed a mathematical combination and is computed as:

$$\binom{N}{n} = \frac{N(N-1)...(N-N+1)}{n!} = \frac{N!}{n!(N-n)!}$$

where a factorial, n! is n*(n-1)*(n-2)...(1) and zero factorial (0!) is one by convention. The number of possible samples with exactly x defectives is the combination associated with obtaining x defectives from m possible defective items and n-x good items from N-m good items:

$$\binom{m}{x}\binom{N-m}{n-x} = \frac{m!}{x!(m-x)!} = \frac{m!}{x!(m-x)!} \times \frac{(N-m)!}{(n-x)!(N-m-n+x)!}$$

Given these possible numbers of samples, the probability of having exactly x defective items in the sample is given by the ratio as the hypergeometric series:

$$P(X-x) = \frac{\binom{m}{x}\binom{N-m}{n-x}}{\binom{N}{n}}$$

With this function, we can calculate the probability of obtaining different numbers of defectives in a sample of a given size. Suppose that the actual fraction of defectives in the lot is p and the actual fraction of no defectives is q, then p plus q is one, resulting in m = Np, and N-m = Nq. Then, a function g(p) representing the probability of having r or less defective items in a sample of size n is obtained by substituting m and N into Equation and summing over the acceptable defective number of items:

$$g(p)=\sum_{x=0}^{r}P(X=x)=\sum_{x=0}^{r}\frac{\binom{Np}{x}\binom{Nq}{n-x}}{\binom{N}{n}}$$

If the number of items in the lot, N, is large in comparison with the sample size n, then the function g(p) can be approximated by the binomial distribution:

$$g(p)=1-\sum_{x=r+1}^{r}\binom{n}{x}p^{x}q^{n-x}$$

or,

$$g(p)=1-\sum_{x=r+1}^{r}\binom{n}{x}p^{x}q^{n-x}$$

The function g(p) indicates the probability of accepting a lot, given the sample size n and the number of allowable defective items in the sample r. The function g(p) can be represented graphical for each combination of sample size n and number of allowable defective items r, as shown in Figure 8.1. Each curve is referred to as the operating characteristic curve (OC curve) in this graph.

For the special case of a single sample (n=1), the function g(p) can be simplified:

$$g(p)=\binom{1}{0}p^{0}q^{1}=q$$

so that the probability of accepting a lot is equal to the fraction of acceptable items in the lot.

For example, there is a probability of 0.5 that the lot may be accepted from a single sample test even if fifty per cent of the lot is defective.For any combination of n and r, we can read off the value of g(p) for a given p from the corresponding OC curve. For example, n = 15 is specified in Figure 8.1.

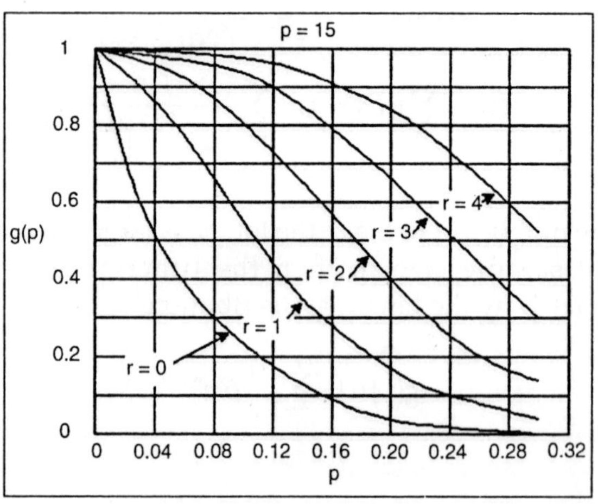

Fig. 8.1 Example Operating Characteristic Curves Indicating Probability of Lot Acceptance

Then, for various values of r, we find:

r=0	p=24%	g(p) ≈ 2%
r=0	p=4%	g(p) ≈ 54%
r=1	p=24%	g(p) ≈ 10%
r=1	p=4%	g(p) ≈ 88%

The producer's and consumer's risk can be related to various points on an operating characteristic curve. Producer's risk is the chance that otherwise *acceptable* lots fail the sampling plan (ie. have more than the allowable number of defective items in the sample) solely due to random fluctuations in the selection of the sample. In contrast, consumer's risk is the chance that an unacceptable lot is acceptable (ie. has less than the allowable number of defective items in the sample) due to a better than average quality in the sample.

For example, suppose that a sample size of 15 is chosen with a trigger level for rejection of one item. With a four per cent acceptable level and a greater than four per cent defective fraction, the consumer's risk is at most eighty-eight per cent. In contrast, with a four per cent acceptable level and a four per cent defective fraction, the producer's risk is at most 1-0.88 =

0.12 or twelve per cent. In specifying the sampling plan implicit in the operating characteristic curve, the supplier and consumer of materials or work must agree on the levels of risk acceptable to themselves.

If the lot is of acceptable quality, the supplier would like to minimize the chance or risk that a lot is rejected solely on the basis of a lower than average quality sample. Similarly, the consumer would like to minimize the risk of accepting under the sampling plan a deficient lot. In addition, both parties presumably would like to minimize the costs and delays associated with testing.

Devising an acceptable sampling plan requires trade off the objectives of risk minimization among the parties involved and the cost of testing.

ACCEPTANCE PROBABILITY CALCULATION

Suppose that the sample size is five (n=5) from a lot of one hundred items (N=100). The lot of materials is to be rejected if any of the five samples is defective (r = 0). In this case, the probability of acceptance as a function of the actual number of defective items can be computed by noting that for r = 0, only one term (x = 0) need be considered in Equation.

Thus, for N = 100 and n = 5:

$$g(p) = \frac{\binom{100p}{0}\binom{100q}{5}}{\binom{100}{5}}$$

For a two per cent defective fraction (p = 0.02), the resulting acceptance value is:

$$g(p) = \frac{\binom{2}{0}\binom{98}{5}}{\binom{100}{5}} = \frac{\dfrac{98!}{93 \sim 5!}}{\dfrac{100!}{95!5!}} = \frac{98!95!}{93!100!} = 0.9020$$

Using the binomial approximation in Equation the comparable calculation would be:

$$g(p) \approx \binom{5}{0} p^0 q^5 = q^5 = (0.98)^5 = 0.9039$$

which is a difference of 0.0019, or 0.21 per cent from the actual value of 0.9020 found above.

If the acceptable defective proportion was two per cent (so $p_1 = p_2 = 0.02$), then the chance of an incorrect rejection (or producer's risk) is 1-g(0.02) = 1-0.9 = 0.1 or ten per cent. Note that a prudent producer should insure better than minimum quality products to reduce the probability or chance of rejection under this sampling plan. If the actual proportion of defectives was one per cent, then the producer's risk would be only five per cent with this sampling plan.

Designing a Sampling Plan

Suppose that an owner (or product "consumer" in the terminology of quality control) wishes to have zero defective items in a facility with 5,000 items of a particular kind. What would be the different amounts of consumer's risk for different sampling plans? With an acceptable quality level of no defective items (so $p_1 = 0$), the allowable defective items in the sample is zero (so $r = 0$) in the sampling plan. Using the binomial approximation, the probability of accepting the 5,000 items as a function of the fraction of actual defective items and the sample size is:

$$g(p) = (1 - p)^n$$

To insure a ninety per cent chance of rejecting a lot with an actual percentage defective of one per cent ($p = 0.01$), the required sample size would be calculated as:

$$g(p) = 1 - 0.90 = 0.1 - (1 - 0.01)$$

Then,

$$n = \frac{\ln(0.1)}{\ln(0.99)} = \frac{-230}{-0.01} \approx 229$$

As can be seen, large sample sizes are required to insure relatively large probabilities of zero defective items.

STATISTICAL QUALITY CONTROL WITH SAMPLING BY VARIABLES

As described in the previous section, sampling by attributes is based on a classification of items as *good* or *defective*. Many work and material attributes possess continuous properties, such as strength, density or length. With the sampling by attributes procedure, a particular level of a variable quantity must be defined as acceptable quality. More generally, two items classified as *good* might have quite different strengths or other attributes. Intuitively, it seems reasonable that some "credit" should be provided for exceptionally good items in a sample.

Sampling by variables was developed for application to continuously measurable quantities of this type. The procedure uses measured values of an attribute in a sample to determine the overall acceptability of a batch or lot. Sampling by variables has the advantage of using more information from tests since it is based on actual measured values rather than a simple classification. As a result, acceptance sampling by variables can be more efficient than sampling by attributes in the sense that fewer samples are required to obtain a desired level of quality control.

In applying sampling by variables, an acceptable lot quality can be defined with respect to an upper limit U, a lower limit L, or both. With these boundary conditions, an acceptable quality level can be defined as a maximum allowable fraction of defective items, M. In Figure 8.2, the probability distribution of item attribute x is illustrated. With an upper limit U, the fraction of defective items is equal to the area under the distribution function to the right of U (so that x £ U). This fraction of defective items would be compared to the allowable fraction M to determine the acceptability of a lot. With both a lower and an upper limit on acceptable quality, the fraction defective would be the fraction of items greater than the upper limit or less than the lower limit. Alternatively, the limits could be imposed upon the acceptable *average* level of the variable

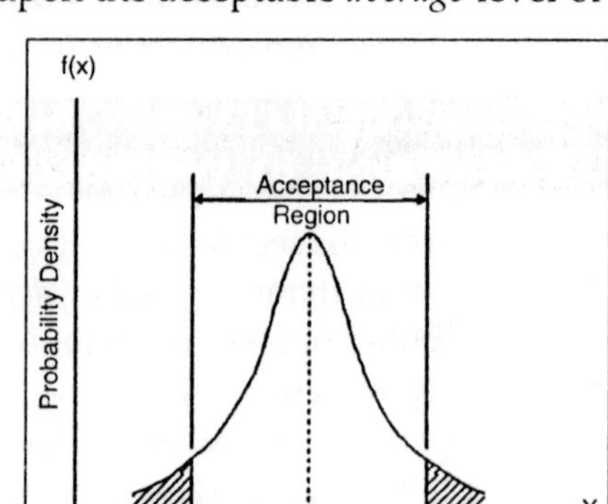

Fig. 8.2 Variable Probability Distributions and Acceptance Regions

In sampling by variables, the fraction of defective items is estimated by using measured values from a sample of items. As with sampling by attributes, the procedure assumes a random sample of a give size is obtained from a lot or batch. In the application of sampling by variables plans, the measured characteristic is virtually always assumed to be *normally distributed* as illustrated in Figure 8.2. The normal distribution is likely to be a reasonably good assumption for many measured characteristics such as material density or degree of soil compaction.

The Central Limit Theorem provides a general support for the assumption: if the source of variations is a large number of small and independent random effects, then the resulting distribution of values will approximate the normal distribution. If the distribution of measured values is not likely to be approximately normal, then sampling by attributes should be adopted. Deviations from normal distributions may appear as *skewed* or non-symmetric distributions, or as distributions with fixed upper and lower limits.

The fraction of defective items in a sample or the chance that the population average has different values is estimated from two statistics obtained from the sample: the sample mean and standard deviation. Mathematically, let n be the number of items in the sample and x_i, i = 1,2,3...,n, be the measured values of the variable characteristic x.

Then an estimate of the overall population mean μ *is the sample mean* $\overline{x}$:

$$\mu \approx \overline{x} = \frac{1}{n}\sum_{i=1}^{n} x_i$$

An estimate of the population standard deviation is s, the square root of the sample variance statistic:

$$\sigma^2 \approx s^2 = \frac{1}{n-1}\sum_{i=1}^{n}(x_i - \overline{x}) = \frac{1}{n-1}\left(\sum_{i=1}^{n} x_i^2 - n\overline{x}^{2}\right)$$

Based on these two estimated parameters and the desired limits, the various fractions of interest for the population can be calculated.

The probability that the average value of a population is greater than a particular lower limit is calculated from the test statistic:

$$t_L = \frac{\overline{x} - L}{s/\sqrt{n}} = \frac{(\overline{x} - L)\sqrt{n}}{s}$$

which is t-distributed with n-1 degrees of freedom. If the population standard deviation is known in advance, then this known value is substituted for the estimate sand the resulting test statistic would be normally distributed. The distribution is similar in appearance to a standard normal distribution, although the spread or variability in the function *decreases* as the degrees of freedom parameter *increases*. As the number of degrees of freedom becomes very large, the t-distribution coincides with the normal distribution.

With an upper limit, the calculations are similar, and the probability that the average value of a population is less than a particular upper limit can be calculated from the test statistic:

$$t_U = \frac{U - \overline{x}}{s/\sqrt{n}} = \frac{(U - \overline{x})\sqrt{n}}{s}$$

With both upper and lower limits, the sum of the

probabilities of being above the upper limit or below the lower limit can be calculated. The calculations to estimate the fraction of items above an upper limit or below a lower limit are very similar to those for the population average.

The only difference is that the square root of the number of samples does not appear in the test statistic formulas:

$$t_{AL} = \frac{\bar{x} - L}{s}$$

and,

$$t_{AU} = \frac{U - \bar{x}}{s}$$

where t_{AL} is the test statistic for all items with a lower limit and t_{AU} is the test statistic for all items with a upper limit. For example, the test statistic for items above an upper limit of 5.5 with $\bar{x} = 4.0$, $s = 3.0$, and $n = 5$ is $t_{AU} = (8.5\text{-}4.0)/3.0 = 1.5$ with $n-1 = 4$ degrees of freedom. Instead of using sampling plans that specify an allowable fraction of defective items, it saves computations to simply write specifications in terms of the allowable test statistic values themselves. This procedure is equivalent to requiring that the sample average be at least a pre-specified number of standard deviations away from an upper or lower limit. For example, with $\bar{x} = 4.0$, $U = 8.5$, $s = 3.0$ and $n = 41$, the sample mean is only about $(8.5\text{-}4.0)/3.0 = 1.5$ standard deviations away from the upper limit.

To summarize, the application of sampling by variables requires the specification of a sample size, the relevant upper or limits, and either:

- The allowable fraction of items falling outside the designated limits or
- The allowable probability that the population average falls outside the designated limit. Random samples are drawn from a pre-defined population and tested to obtained measured values of a variable attribute. From these measurements, the sample mean, standard deviation, and quality control test statistic are calculated.

Finally, the test statistic is compared to the allowable trigger level and the lot is either accepted or rejected. It is also possible to apply sequential sampling in this procedure, so that a batch may be subjected to additional sampling and testing to further refine the test statistic values. With sampling by variables, it is notable that a producer of material or work can adopt two general strategies for meeting the required specifications. First, a producer may insure that the average quality level is quite high, even if the variability among items is high. This strategy is illustrated in Figure 8.3 as a "high quality average" strategy. Second, a producer may meet a desired quality target by reducing the *variability* within each batch. In Figure 8.3, this is labeled the "low variability" strategy. In either case, a producer should maintain high standards to avoid rejection of a batch.

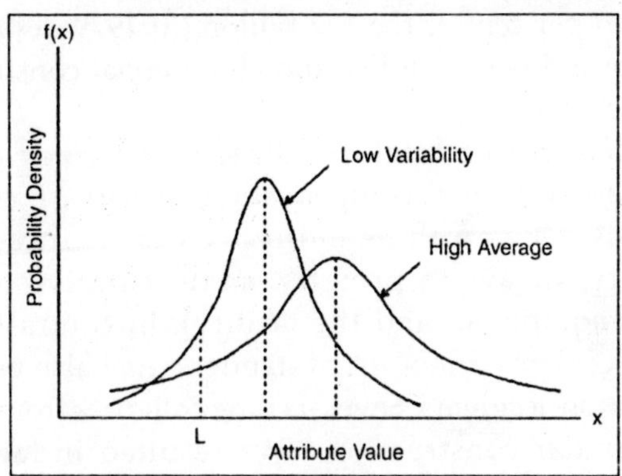

Fig. 8.3 Testing for Defective Component Strengths

Testing for defective component strengths

Suppose that an inspector takes eight strength measurements with the following results:

4.3, 4.8, 4.6, 4.7, 4.4, 4.6, 4.7, 4.6

In this case, the sample mean and standard deviation can be calculated using Equations:

$$\bar{x} = 1/8(4.3 + 4.8 + 4.6 + 4.7 + 4.4 + 4.6 + 4.7 + 4.6) = 4.59$$

$$s^2=[1/(8-1)][(4.3-4.59)^2 + (4.8-4.59)^2 + (4.6-4.59)^2 + (4.7-4.59)^2 +$$

$(4.4\text{-}4.59)^2 + (4.6\text{-}4.59)^2 + (4.7\text{-}4.59)^2 + (4.6\text{-}4.59)^2] = 0.16$

The percentage of items below a lower quality limit of L = 4.3 is estimated from the test statistic t_{AL} *in Equation:*

$$t_{AU} = \frac{4.59 - 4.3}{0.16} = 1.81$$

SAFETY

Construction is a relatively hazardous undertaking. As Table illustrates, there are significantly more injuries and lost workdays due to injuries or illnesses in construction than in virtually any other industry. These work related injuries and illnesses are exceedingly costly. The *Construction Industry Cost Effectiveness Project* estimated that accidents cost 8.9 billion or nearly seven per cent of the 137 billion (in 1979 dollars) spent annually for industrial, utility and commercial construction in the United States.

Included in this total are direct costs (medical costs, premiums for workers' compensation benefits, liability and property losses) as well as indirect costs (reduced worker productivity, delays in projects, administrative time, and damage to equipment and the facility). In contrast to most industrial accidents, innocent bystanders may also be injuried by construction accidents. Several crane collapses from high rise buildings under construction have resulted in fatalities to passerbys. Prudent project managers and owners would like to reduce accidents, injuries and illnesses as much as possible.

Table. Nonfatal Occupational Injury and Illness Incidence Rates

Industry	1996	2006
Agriculture, forestry, fishing	8.7	6
Mining	5.4	3.5
Construction	9.9	5.9
Manufacturing	10.6	6
Trade,Transportation and utilities	8.7	5
Financial activities	2.4	1.5
Professional and business services	6.0	1.2

As with all the other costs of construction, it is a mistake for owners to ignore a significant category of costs such as injury and illnesses. While contractors may pay insurance premiums directly, these costs are reflected in bid prices or contract amounts. Delays caused by injuries and illnesses can present significant opportunity costs to owners. In the long run, the owners of constructed facilities must pay all the costs of construction. For the case of injuries and illnesses, this general principle might be slightly qualified since significant costs are borne by workers themselves or society at large. However, court judgments and insurance payments compensate for individual losses and are ultimately borne by the owners.

The causes of injuries in construction are numerous. Table lists the reported causes of accidents in the US construction industry in 1997 and 2004. A similar catalogue of causes would exist for other countries. The largest single category for both injuries and fatalities are individual falls. Handling goods and transportation are also a significant cause of injuries. From a management perspective, however, these reported causes do not really provide a useful prescription for safety policies. An individual fall may be caused by a series of coincidences: a railing might not be secure, a worker might be inattentive, the footing may be slippery, etc. Removing any one of these compound causes might serve to prevent any particular accident. However, it is clear that conditions such as unsecured railings will normally increase the risk of accidents. Table provides a more detailed list of causes of fatalities for construction sites alone, but again each fatality may have multiple causes.

Table. Fatal Occupational Injuries in Construction

Year	**1997**	**2004**
Total fatalities	1,107	1,234
Falls	376	445
Transportation incidents	288	287
Contact with objects & equipment	199	267
Exposure to harmful substances and environments	188	170

Table. Fatality Causes in Construction, 1996/1997 and 2006/2007

Year	96/97	06/07
Total accidents	287	241
Falls from a height	88	45
Struck by a moving vehicle	43	30
Struck by moving/falling object	57	40
Trapped by something overturning/collapsing	16	19
Drowning/asphyxiation	9	16

Various measures are available to improve jobsite safety in construction. Several of the most important occur before construction is undertaken. These include design, choice of technology and education. By altering facility designs, particular structures can be safer or more hazardous to construct. For example, parapets can be designed to appropriate heights for construction worker safety, rather than the minimum height required by building codes.

Choice of technology can also be critical in determining the safety of a jobsite. Safeguards built into machinery can notify operators of problems or prevent injuries. For example, simple switches can prevent equipment from being operating when protective shields are not in place. With the availability of on-board electronics (including computer chips) and sensors, the possibilities for sophisticated machine controllers and monitors has greatly expanded for construction equipment and tools. Materials and work process choices also influence the safety of construction. For example, substitution of alternative materials for asbestos can reduce or eliminate the prospects of long term illnesses such as *asbestiosis*.

Educating workers and managers in proper procedures and hazards can have a direct impact on jobsite safety. The realization of the large costs involved in construction injuries and illnesses provides a considerable motivation for awareness and education. Regular safety inspections and safety meetings have become standard practices on most job sites. Pre-qualification of contractors and sub-contractors with regard to safety is another important avenue for safety improvement. If contractors are only invitied to bid or enter negotiations if they have an acceptable record of safety (as well as quality performance), then a direct incentive is provided to insure adequate safety on the part of contractors.

During the construction process itself, the most important safety related measures are to insure vigilance and cooperation on the part of managers, inspectors and workers. Vigilance involves considering the risks of different working practices. In also involves maintaining temporary physical safeguards such as barricades, braces, guylines, railings, toeboards and the like.

Sets of standard practices are also important, such as:

- Requiring hard hats on site.
- Requiring eye protection on site.
- Requiring hearing protection near loud equipment.
- Insuring safety shoes for workers.
- Providing first-aid supplies and trained personnel on site

While eliminating accidents and work related illnesses is a worthwhile goal, it will never be attained. Construction has a number of characteristics making it inherently hazardous. Large forces are involved in many operations. The jobsite is continually changing as construction proceeds. Workers do not have fixed worksites and must move around a structure under construction. The tenure of a worker on a site is short, so the worker's familiarity and the employer-employee relationship are less settled than in manufacturing settings. Despite these peculiarities and as a result of exactly these special problems, improving worksite safety is a very important project management concern.

Trench collapse

To replace 1,200 feet of a sewer line, a trench of between 12.5 and 18 feet deep was required down the center of a four lane street. The contractor chose to begin excavation of the trench from the shallower end, requiring a 12.5 deep trench. Initially, the contractor used a nine foot high, four foot wide steel trench box for soil support. A trench box is a rigid steel frame consisting of two walls supported by welded struts with open sides and ends. This method had the advantage that traffic could be maintained in at least two lanes during the reconstruction work.

In the shallow parts of the trench, the trench box seemed to adequately support the excavation. However, as the trench got deeper, more soil was unsupported below the trench box. Intermittent soil collapses in the trench began to occur. Eventually, an old parallel six inch water main collapsed, thereby saturating the soil and leading to massive soil collapse at the bottom of the trench. Replacement of the water main was added to the initial contract. At this point, the contractor began sloping the sides of the trench, thereby requiring the closure of the entire street.

The initial use of the trench box was convenient, but it was clearly inadequate and unsafe. Workers in the trench were in continuing danger of accidents stemming from soil collapse. Disruption to surrounding facilities such as the parallel water main was highly likely. Adoption of a tongue and groove vertical sheeting system over the full height of the trench or, alternatively, the sloping excavation eventually adopted are clearly preferable.

9

Cost Control

THE COST CONTROL PROBLEM

During the execution of a project, procedures for project control and record keeping become indispensable tools to managers and other participants in the construction process. These tools serve the dual purpose of recording the financial transactions that occur as well as giving managers an indication of the progress and problems associated with a project. The problems of project control are aptly summed up in an old definition of a project as "any collection of vaguely related activities that are ninety per cent complete, over budget and late." The task of project control systems is to give a fair indication of the existence and the extent of such problems.

In this chapter, we consider the problems associated with resource utilization, accounting, monitoring and control during a project. In this discussion, we emphasize the project management uses of accounting information. Interpretation of project accounts is generally not straightforward until a project is completed, and then it is too late to influence project management. Even after completion of a project, the accounting results may be confusing. Hence, managers need to know how to interpret accounting information for the purpose of project management. In the process of considering management problems, however, we shall discuss some of the common accounting systems and conventions, although our purpose is not to provide a comprehensive survey of accounting procedures. The limited objective of project control deserves emphasis. Project control procedures are primarily intended to identify deviations from the project plan rather than to suggest possible areas for cost savings. This characteristic reflects the advanced stage at which project control becomes important. The time at which major cost savings can be achieved is during planning and design for the project. During the actual construction, changes are likely to delay the project and lead to inordinate cost increases. As a result, the focus of project control is on fulfilling the original design plans or indicating deviations from these plans, rather than on searching for significant improvements and cost savings. It is only when a rescue operation is required that major changes will normally occur in the construction plan.

Finally, the issues associated with integration of information will require some discussion. Project management activities and functional concerns are intimately linked, yet the techniques used in many instances do not facilitate comprehensive or integrated consideration of project activities. For example, schedule information and cost accounts are usually kept separately. As a result, project managers themselves must synthesize a comprehensive view from the different reports on the project plus their own field observations. In particular, managers are often forced to infer the cost impacts of schedule changes, rather than being provided with aids for this process. Communication or integration of various types of information can serve a number of useful purposes, although it does require special attention in the establishment of project control procedures.

THE PROJECT BUDGET

For cost control on a project, the construction plan and the associated cash flow estimates can provide the baseline reference for subsequent project monitoring and control. For schedules, progress on individual activities and the achievement of milestone completions can be compared with the project schedule to monitor the progress of activities. Contract and job specifications provide the criteria by which to assess and assure the required quality of construction. The final or detailed cost estimate provides a baseline for the assessment of financial performance during the project. To the extent that costs are within the detailed cost estimate, then the project is thought to be under *financial control*. Overruns in particular cost categories signal the possibility of problems and give an indication of exactly what problems are being encountered. Expense oriented construction planning and control focuses upon the categories included in the final cost estimation. This focus is particular relevant for projects with few activities and considerable repetition such as grading and paving roadways.

For control and monitoring purposes, the original detailed cost estimate is typically converted to a *project budget*, and the project budget is used subsequently as a guide for management. Specific items in the detailed cost estimate become job cost elements. Expenses incurred during the course of a project are recorded in specific job cost accounts to be compared with the original cost estimates in each category. Thus, individual job cost accounts generally represent the basic unit for cost control. Alternatively, job cost accounts may be disaggregated or divided into *work elements* which are related both to particular scheduled activities and to particular cost accounts. Work element divisions will be described in Section 12.8. In addition to cost amounts, information on material quantities and labor inputs within each job account is also typically retained in the project budget. With this information, actual materials usage and labor employed can be compared to the expected requirements. As a result, cost overruns or savings on particular items can be identified as due to changes in unit prices, labor productivity or in the amount of material consumed.

The number of cost accounts associated with a particular project can vary considerably. For constructors, on the order of four hundred separate cost accounts might be used on a small project. These accounts record all the transactions associated with a project. Thus, separate accounts might exist for different types of materials, equipment use, payroll, project office, etc. Both physical and non-physical resources are represented, including overhead items such as computer use or interest charges. Table summarizes a typical set of cost accounts that might be used in building construction. Note that this set of accounts is organized hierarchically, with seven major divisions (accounts 201 to 207) and numerous sub-divisions under each division. This hierarchical structure facilitates aggregation of costs into pre-defined categories; for example, costs associated with the superstructure (account 204) would be the sum of the underlying subdivisions (ie. 204.1, 204.2, etc.) or finer levels of detail.

Illustrative Set of Project Cost Accounts:

- 201: Clearing and Preparing Site
- 202: Substructure
- 202.1: Excavation and Shoring
- 202.2: Piling
- 202.3: Concrete Masonry
- 202.31: Mixing and Placing
- 202.32: Formwork
- 202.33: Reinforcing
- 203: Outside Utilities (water, gas, sewer, etc.)
- 204: Superstructure
- 204.1: Masonry Construction
- 204.2: Structural Steel
- 204.3: Wood Framing, Partitions, etc.
- 204.4: Exterior Finishes (brickwork, terra cotta, cut stone, etc.)
- 204.5: Roofing, Drains, Gutters, Flashing, etc.
- 204.6: Interior Finish and Trim
- 204.61: Finish Flooring, Stairs, Doors, Trim
- 204.62: Glass, Windows, Glazing
- 204.63: Marble, Tile, Terrazzo
- 204.64: Lathing and Plastering
- 204.65: Soundproofing and Insulation
- 204.66: Finish Hardware
- 204.67: Painting and Decorating
- 204.68: Waterproofing

- 204.69: Sprinklers and Fire Protection
- 204.7: Service Work
- 204.71: Electrical Work
- 204.72: Heating and Ventilating
- 204.73: Plumbing and Sewage
- 204.74: Air Conditioning
- 204.72: Fire Alarm, Telephone, Security, Miscellaneous
- 205: Paving, Curbs, Walks
- 206: Installed Equipment (elevators, revolving doors, mailchutes, etc.)
- 207: Fencing

In developing or implementing a system of cost accounts, an appropriate numbering or coding system is essential to facilitate communication of information and proper aggregation of cost information. Particular cost accounts are used to indicate the expenditures associated with specific projects and to indicate the expenditures on particular items throughout an organization. These are examples of different *perspectives* on the same information, in which the same information may be summarized in different ways for specific purposes. Thus, more than one aggregation of the cost information and more than one application programme can use a particular cost account. Separate identifiers of the type of cost account and the specific project must be provided for project cost accounts or for financial transactions. Converting a final cost estimate into a project budget compatible with an organization's cost accounts is not always a straightforward task.

For example, labor and material quantities might be included for each of several physical components of a project. For cost accounting purposes, labor and material quantities are aggregated by type no matter for which physical component they are employed. For example, particular types of workers or materials might be used on numerous different physical components of a facility. Moreover, the categories of cost accounts established within an organization may bear little resemblance to the quantities included in a final cost estimate. This is particularly true when final cost estimates are prepared in accordance with an external reporting requirement rather than in view of the existing cost accounts within an organization.

One particular problem in forming a project budget in terms of cost accounts is the treatment of contingency amounts. These allowances are included in project cost estimates to accommodate unforeseen events and the resulting costs. However, in advance of project completion, the source of contingency expenses is not known. Realistically, a budget accounting item for *contingency allowance* should be established whenever a contingency amount was included in the final cost estimate. A second problem in forming a project budget is the treatment of inflation. Typically, final cost estimates are formed in terms of real dollars and an item reflecting inflation costs is added on as a percentage or lump sum. This inflation allowance would then be allocated to individual cost items in relation to the actual expected inflation over the period for which costs will be incurred.

Project Budget for a Design Office

An example of a small project budget is shown in Table 12-2. This budget might be used by a design firm for a specific design project. While this budget might represent all the work for this firm on the project, numerous other organizations would be involved with their own budgets. In Table, a summary budget is shown as well as a detailed listing of costs for individuals in the Engineering Division. For the purpose of consistency with cost accounts and managerial control, labor costs are aggregated into three groups: the engineering, architectural and environmental divisions. The detailed budget shown in Table applies only to the engineering division labor; other detailed budgets amounts for categories such as supplies and the other work divisions would also be prepared.

Note that the salary costs associated with individuals are aggregated to obtain the total labor costs in the engineering group for the project. To perform this aggregation, some means of identifying individuals within organizational groups is required. Accompanying a budget of this nature, some estimate of the actual man-hours of labor required by project task would also be prepared. Finally, this budget might be used for internal purposes alone. In submitting financial bills and reports to the client, overhead and contingency amounts might be combined with the direct labor costs to establish an aggregate billing rate per hour. In this case, the overhead, contingency and profit would represent *allocated costs* based on the direct labor costs.

Table. Example of a Small Project Budget for a Design Firm

	Budget Summary
Personnel:	
Architectural Division	$ 67,251.00
Engineering	45,372.00
Environmental Division	28,235.00
Total	$140,858.00
Other Direct Expenses:	
Travel	2,400.00
Supplies	1,500.00
Communication	600.00
Computer Services	1,200.00
Total	$ 5,700.00
Overhead	$ 175,869.60
Contingency and Profit	$ 95,700.00
Total	$ 418,127.60
	Engineering Personnel Detail
Senior Engineer	
Associate Engineer	$ 11,562.00
Engineer Technician	21,365.00
Total	12,654.00
	$ 45,372.00

Project Budget for a Constructor

Table illustrates a summary budget for a constructor. This budget is developed from a project to construct a wharf. As with the example design office budget above, costs are divided into direct and indirect expenses. Within direct costs, expenses are divided into material, subcontract, temporary work and machinery costs. This budget indicates aggregate amounts for the various categories. Cost details associated with particular cost accounts would supplement and support the aggregate budget shown in Table. A profit and a contingency amount might be added to the basic budget of $1,715,147 shown in Table for completeness.

Table. An Example of a Project Budget for a Wharf Project (Amounts in Thousands of Dollars)

	Material Cost	Subcontract Work	Temporary Work	Machinery Cost	Total Cost
Steel Piling	$292,172	$129,178	$16,389	$0	$437,739
Tie-rod	88,233	29,254	0	0	117,487
Anchor-Wall	130,281	60,873	0	0	191,154
Backfill	242,230	27,919	0	0	300,149
Coping	42,880	22,307	13,171	0	78,358
Dredging	0	111,650	0	0	111,650
Fender	48,996	10,344	0	1,750	61,090
Other	5,000	32,250	0	0	37,250
Sub-total	$849,800	$423,775	$29,560	$1,750	$1,304,885
			Summary		
			Total of direct cost		$1,304,885
			Indirect Cost		
			Common Temporary Work		19,320
			Common Machinery		80,934
			Transportation		15,550
			Office Operating Costs		294,458
			Total of Indirect Cost		410,262
			Total Project Cost		$1,715,147

FORECASTING FOR ACTIVITY COST CONTROL

For the purpose of project management and control, it is not sufficient to consider only the past record of costs and revenues incurred in a project. Good managers should focus upon future revenues, future costs and technical problems. For this purpose, traditional financial accounting schemes are not adequate to reflect the dynamic nature of a project. Accounts typically focus on recording routine costs and past expenditures associated with activities. Generally, past expenditures represent *sunk costs* that cannot be altered in the future and may or may not be relevant in the future. For example, after the completion of some activity, it may be discovered that some quality flaw renders the work useless. Unfortunately, the resources expended on the flawed construction will generally be *sunk* and cannot be recovered for re-construction (although it may be possible to change the burden of who pays for these resources by financial withholding or charges; owners will typically attempt to have constructors or designers pay for changes due to quality flaws). Since financial accounts are historical in nature, some means of forecasting or projecting the future course of a project is essential for management control. In this section, some methods for cost control and simple forecasts are described.

An example of forecasting used to assess the project status is shown in Table.

In this example, costs are reported in five categories, representing the sum of all the various cost accounts associated with each category:

- *Budgeted Cost*: The budgeted cost is derived from the detailed cost estimate prepared at the start of the project. Examples of project budgets were presented in Section. The factors of cost would be referenced by cost account and by a prose description.
- *Estimated total cost*: The estimated or forecast total cost in each category is the current best estimate of costs based on progress and any changes since the budget was formed. Estimated total costs are the sum of cost to date, commitments and exposure. Methods for estimating total costs are described below.

- *Cost Committed and Cost Exposure*: Estimated cost to completion in each category in divided into firm commitments and estimated additional cost or*exposure*. Commitments may represent material orders or subcontracts for which firm dollar amounts have been committed.
- *Cost to Date*: The actual cost incurred to date is recorded in column 6 and can be derived from the financial record keeping accounts.
- *Over or (Under)*: A final column in Table indicates the amount over or under the budget for each category. This column is an indicator of the extent of variance from the project budget; items with unusually large overruns would represent a particular managerial concern. Note that *variance* is used in the terminology of project control to indicate a difference between budgeted and actual expenditures. The term is defined and used quite differently in statistics or mathematical analysis. In Table, labor costs are running higher than expected, whereas subcontracts are less than expected.

The current status of the project is a forecast budget overrun of $5,950. with 23 per cent of the budgeted project costs incurred to date.

Table. Illustration of a Job Status Report

Factor	Budgeted Cost	Estimated Total Cost	Cost Committed	Cost Exposure	Cost To Date	Over or (Under)
Labor	$99,406	$102,342	$49,596	---	$52,746	$2,936
Material	88,499	88,499	42,506	45,993	---	0
Subcontracts	198,458	196,323	83,352	97,832	15,139	(2,135)
Equipment	37,543	37,543	23,623	---	13,920	0
Other	72,693	81,432	49,356	---	32,076	8,739
Total	496,509	506,139	248,433	143,825	113,881	5,950

For project control, managers would focus particular attention on items indicating substantial deviation from budgeted amounts. In particular, the cost overruns in the labor and in the "other expense category would be worthy of attention by a project manager in Table. A next step would be to look in greater detail at the various components of these categories. Overruns in cost might be due to lower than expected productivity, higher than expected wage rates, higher than expected material costs, or other factors.

Even further, low productivity might be caused by inadequate training, lack of required resources such as equipment or tools, or inordinate amounts of re-work to correct quality problems. Review of a job status report is only the first step in project control.

The job status report illustrated in Table employs explicit estimates of ultimate cost in each category of expense. These estimates are used to identify the actual progress and status of a expense category. Estimates might be made from simple linear extrapolations of the productivity or cost of the work to date on each project item. Algebraically, a linear estimation formula is generally one of two forms.

Using a linear extrapolation of costs, the forecast total cost, C_f, is:

$$C_f = \frac{C_t}{p_t}$$

where C_t is the cost incurred to time t and p_t is the proportion of the activity completed at time t. For example, an activity which is 50 per cent complete with a cost of $40,000 would be estimated to have a total cost of $40,000/0.5 = $80,000. More elaborate methods of forecasting costs would disaggregate costs into different categories, with the total cost the sum of the forecast costs in each category. *Alternatively, the use of measured unit cost amounts can be used for forecasting total cost.*

The basic formula for forecasting cost from unit costs is:

$$C_f = Wc_t$$

where C_f is the forecast total cost, W is the total units of work, and c_t is the average cost per unit of work experienced up to time t. If the average unit cost is $50 per unit of work on a particular activity and 1,600 units of work exist, then the expected cost is (1,600)(50) = \$80,000 for completion. The unit cost in Equation may be replaced with the hourly productivity and the unit cost per hour (or other appropriate time period), resulting in the equation:

$$C_f = Wh_t u_t$$

where the cost per work unit (c_t) is replaced by the time per unit, h_t, divided by the cost per unit of time, u_t.

More elaborate forecasting systems might recognize peculiar problems associated with work on particular items and modify these simple proportional cost estimates. For example, if productivity is improving as workers and managers become more familiar with the project activities, the estimate of total costs for an item might be revised downward.

In this case, the estimating equation would become:

$$C_f = C_t + (W - W_t)C_t$$

where forecast total cost, C_f, is the sum of cost incurred to date, C_t, and the cost resulting from the remaining work (W-W_t) multiplied by the expected cost per unit time period for the remainder of the activity, c_t.

If the project manager was assured that the improved productivity could be maintained for the remainder of the project (consisting of 800 units of work out of a total of 1600 units of work), the cost estimate would be (50)(800) + (45)(800) = \$76,000 for completion of the activity. Note that this forecast uses the actual average productivity achieved on the first 800 units and uses a forecast of productivity for the remaining work. Historical changes in productivity might also be used to represent this type of non-linear changes in work productivity on particular activities over time.

In addition to changes in productivities, other components of the estimating formula can be adjusted or more detailed estimates substituted. For example, the change in unit prices due to new labor contracts or material supplier's prices might be reflected in estimating future expenditures. In essence, the same problems encountered in preparing the detailed cost estimate are faced in the process of preparing exposure estimates, although the number and extent of uncertainties in the project environment decline as work progresses. The only exception to this rule is the danger of quality problems in completed work which would require re-construction.

Each of the estimating methods described above require current information on the state of work accomplishment for particular activities.

There are several possible methods to develop such estimates, including:

- *Units of Work Completed*: For easily measured quantities the actual proportion of completed work amounts can be measured. For example, the linear feet of piping installed can be compared to the required amount of piping to estimate the percentage of piping work completed.
- *Incremental Milestones*: Particular activities can be sub-divided or "decomposed" into a series of milestones, and the milestones can be used to indicate the percentage of work complete based on historical averages. For example, the work effort involved with installation of standard piping might be divided into four milestones:
  - *Spool in place*: 20% of work and 20% of cumulative work.
  - *Ends welded*: 40% of work and 60% of cumulative work.
  - *Hangars and Trim Complete*: 30% of work and 90% of cumulative work.
  - *Hydrotested and Complete*: 10% of work and 100% of cumulative work.

Thus, a pipe section for which the ends been welded would be reported as 60% complete.

- *Opinion*: Subjective judgments of the percentage complete can be prepared by inspectors, supervisors or project managers themselves. Clearly, this estimated technique can be biased by optimism, pessimism or inaccurate observations. Knowledgeable estimaters and adequate field observations are required to obtain sufficient accuracy with this method.

- *Cost Ratio*: The cost incurred to date can also be used to estimate the work progress. For example, if an activity was budgeted to cost $20,000 and the cost incurred at a particular date was $10,000, then the estimated percentage complete under the cost ratio method would be 10,000/20,000 = 0.5 or fifty per cent. This method provides no independent information on the actual percentage complete or any possible errors in the activity budget: the cost forecast will always be the budgeted amount. Consequently, managers must use the estimated costs to complete an activity derived from the cost ratio method with extreme caution.

Systematic application of these different estimating methods to the various project activities enables calculation of the percentage complete or the productivity estimates used in preparing job status reports.

In some cases, automated data acquisition for work accomplishments might be instituted. For example, transponders might be moved to the new work limits after each day's activity and the new locations automatically computed and compared with project plans. These measurements of actual progress should be stored in a central database and then processed for updating the project schedule.

Estimated Total Cost to Complete an Activity

Suppose that we wish to estimate the total cost to complete piping construction activities on a project. The piping construction involves 1,000 linear feet of piping which has been divided into 50 sections for management convenience. At this time, 400 linear feet of piping has been installed at a cost of $40,000 and 500 man-hours of labor. The original budget estimate was $90,000 with a productivity of one foot per man-hour, a unit cost of $60 per man hour and a total material cost of $ 30,000. Firm commitments of material delivery for the

$30,000 estimated cost have been received. The first task is to estimate the proportion of work completed. Two estimates are readily available. First, 400 linear feet of pipe is in place out of a total of 1000 linear feet, so the proportion of work completed is 400/1000 = 0.4 or 40%.

This is the "units of work completed" estimation method. Second, the cost ratio method would estimate the work complete as the cost-to-date divided by the cost estimate or $40,000/$ 90,000 = 0.44 or 44%. Third, the "incremental milestones" method would be applied by examining each pipe section and estimating a percentage complete and then aggregating to determine the total percentage complete. For example, suppose the following quantities of piping fell into four categories of completeness:

Complete (100%)	380 ft
Hangars And Trim Complete (90%)	20 ft
Ends Welded (60%)	5 ft
Spool In place (20%)	0 ft

Then using the incremental milestones shown above, the estimate of completed work would be 380 + (20)(0.9) + (5)(0.6) + 0 = 401 ft and the proportion complete would be 401 ft/1,000 ft = 0.401 or 40% after rounding.

Once an estimate of work completed is available, then the estimated cost to complete the activity can be calculated. First, a simple linear extrapolation of cost results in an estimate of $40,000/0.4 = $100,000. for the piping construction using the 40% estimate of work completed. This estimate projects a cost overrun of 100,000-90,000 = $10,000.

Second, a linear extrapolation of productivity results in an estimate of (1000 ft.)(500 hrs/400 ft.)($60/hr) + 30,000 = $105,000. for completion of the piping construction. This estimate suggests a variance of 105,000–90,000 = $15,000 above the activity estimate. In making this estimate, labor and material costs entered separately, whereas the two were implicitly combined in the simple linear cost forecast above. The source of the variance can also be identified in this calculation: compared to the original estimate, the labor productivity is 1.25 hours per foot or 25% higher than the original estimate.

Estimated Total Cost for Completion

The forecasting procedures described above assumed linear extrapolations of future costs, based either on the complete experience on the activity or the recent experience. For activities with good historical records, it can be the case that a typically non-linear profile of cost expenditures and completion proportions can be estimated.

Figure 9.1 illustrates one possible non-linear relationships derived from experience in some particular activity. The progress on a new job can be compared to this historical record.

For example, point A in Figure 9.1 suggests a higher expenditure than is normal for the completion proportion. This point represents 40% of work completed with an expenditure of 60% of the budget.

Since the historical record suggests only 50% of the budget should be expended at time of 40% completion, a 60-50 = 10% overrun in cost is expected even if work efficiency can be increased to historical averages. If comparable cost overruns continue to accumulate, then the cost-to-complete will be even higher.

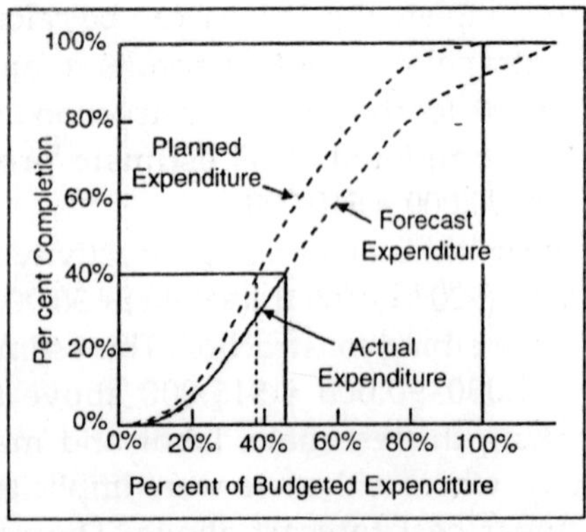

Fig. 9.1 Illustration of Proportion Completion Versus Expenditure for an Activity

FINANCIAL ACCOUNTING SYSTEMS AND COST ACCOUNTS

The cost accounts described in the previous sections provide only one of the various components in a financial accounting system. Before further discussing the use of cost accounts in project control, the relationship of project and financial accounting deserves mention.

Accounting information is generally used for three distinct purposes:

- Internal reporting to project managers for day-to-day planning, monitoring and control.
- Internal reporting to managers for aiding strategic planning.
- External reporting to owners, government, regulators and other outside parties.

External reports are constrained to particular forms and procedures by contractual reporting requirements or by generally accepted accounting practices. Preparation of such external reports is referred to as *financial accounting*. In contrast, *cost* or *managerial* accounting is intended to aid internal managers in their responsibilities of planning, monitoring and control.

Project costs are always included in the system of financial accounts associated with an organization. At the heart of this system, all expense transactions are recorded in a general ledger. The general ledger of accounts forms the basis for management reports on particular projects as well as the financial accounts for an entire organization.

Other components of a financial accounting system include:

- The accounts payable journal is intended to provide records of bills received from vendors, material suppliers, subcontractors and other outside parties. Invoices of charges are recorded in this system as are checks issued in payment. Charges to individual cost accounts are relayed or *posted* to the General Ledger.
- Accounts receivable journals provide the opposite function to that of accounts payable. In this journal,

billings to clients are recorded as well as receipts. Revenues received are relayed to the general ledger.

- Job cost ledgers summarize the charges associated with particular projects, arranged in the various cost accounts used for the project budget.
- Inventory records are maintained to identify the amount of materials available at any time.

In traditional bookkeeping systems, day to day transactions are first recorded in journals. With double-entry bookkeeping, each transaction is recorded as both a debit and a credit to particular accounts in the ledger.

For example, payment of a supplier's bill represents a debit or increase to a project cost account and a credit or reduction to the company's cash account. Periodically, the transaction information is summarized and transferred to ledger accounts. This process is called*posting*, and may be done instantaneously or daily in computerized systems.

In reviewing accounting information, the concepts of *flows* and *stocks* should be kept in mind. Daily transactions typically reflect flows of dollar amounts entering or leaving the organization.

Similarly, use or receipt of particular materials represent flows from or to inventory. An account balance represents the *stock* or cumulative amount of funds resulting from these daily flows. Information on both flows and stocks are needed to give an accurate view of an organization's state. In addition, forecasts of future changes are needed for effective management.

Information from the general ledger is assembled for the organization's financial reports, including balance sheets and income statements for each period. These reports are the basic products of the financial accounting process and are often used to assess the performance of an organization.

Table shows a typical income statement for a small construction firm, indicating a net profit of $ 330,000 after taxes. This statement summarizes the flows of transactions within a year. Table 12-6 shows the comparable balance sheet, indicated a net increase in retained earnings equal to the net profit. The balance sheet reflects the effects of income flows during the year on the overall worth of the organization.

Table. Illustration of an Accounting Statement of Income

Income Statement for the year ended December 31, 19xx	
Gross project revenues	$7,200,000
Direct project costs on contracts	5,500,000
Depreciation of equipment	200,000
Estimating	150,000
Administrative and other expenses	650,000
Subtotal of cost and expenses	6,500,000
Operating Income	700,000
Interest Expense, net	150,000
Income before taxes	550,000
Income tax	220,000
Net income after tax	330,000
Cash dividends	100,000
Retained earnings, current year	230,000
Retention at beginning of year	650,000
Retained earnings at end of year	$880,000.

Table. Illustration of an Accounting Balance Sheet

Balance Sheet December 31, 19xx	
Assets	**Amount**
Cash	$150,000
Payments Receivable	750,000
Work in progress, not claimed	700,000
Work in progress, retention	200,000
Equipment at cost less accumulated depreciation	1,400,000
Total assets	$3,200,000

Assets	Amount
Liabilities and Equity	
Liabilities	
Accounts payable	$950,000
Other items payable (taxes, wages, etc.)	50,000
Long term debts	500,000
Subtotal	1,500,000
Shareholders' funds	
40,000 shares of common stock	
(Including paid-in capital)	820,000
Retained Earnings	880,000
Subtotal	1,700,000
Total Liabilities and Equity	$3,200,000

In the context of private construction firms, particular problems arise in the treatment of uncompleted contracts in financial reports. Under the "completed-contract" method, income is only reported for completed projects. Work on projects underway is only reported on the balance sheet, representing an asset if contract billings exceed costs or a liability if costs exceed billings. When a project is completed, the total net profit (or loss) is reported in the final period as income.

Under the "percentage-of-completion" method, actual costs are reported on the income statement plus a proportion of all project revenues (or billings) equal to the proportion of work completed during the period. The proportion of work completed is computed as the ratio of costs incurred to date and the total estimated cost of the project. Thus, if twenty per cent of a project was completed in a particular period at a direct cost of $180,000 and on a project with expected revenues of $1,000,000, then the contract revenues earned would be calculated as $1,000,000(0.2) = $200,000.

Note that billings and actual receipts might be in excess or less than the calculated revenues of $200,000. On the balance sheet of an organization using the percentage-of-completion method, an asset is usually reported to reflect billings and the estimated or calculated earnings in excess of actual billings.

As another example of the difference in the "percentage-of-completion" and the "completed-contract" methods, consider a three year project to construct a plant with the following cash flow for a contractor:

Year	Contract Expenses	Payments Received
1	$700,000	$900,000
2	180,000	250,000
3	320,000	150,000
Total	$1,200,000	$1,300,000

The supervising architect determines that 60% of the facility is complete in year 1 and 75% in year 2. Under the "percentage-of-completion" method, the net income in year 1 is $780,000 (60% of $1,300,000) less the $700,000 in expenses or $80,000. Under the "completed-contract" method, the entire profit of $100,000 would be reported in year 3.

The "percentage-of-completion" method of reporting period earnings has the advantage of representing the actual estimated earnings in each period. As a result, the income stream and resulting profits are less susceptible to precipitate swings on the completion of a project as can occur with the "completed contract method" of calculating income.

However, the "percentage-of-completion" has the disadvantage of relying upon estimates which can be manipulated to obscure the actual position of a company or which are difficult to reproduce by outside observers. There are also subtleties such as the deferral of all calculated income from a project until a minimum threshold of the project is completed. As a result, interpretation of the income statement and balance sheet of a private organization is not always straightforward. Finally, there are tax disadvantages from using the "percentage-of-completion" method since corporate taxes on expected profits may become due during the project rather than being deferred until the project completion.

As an example of tax implications of the two reporting methods, a study of forty-seven construction firms conducted by the General Accounting Office found that $280 million in taxes were deferred from 1980 to 1984 through use of the "completed-contract" method.

It should be apparent that the "percentage-of-completion" accounting provides only a rough estimate of the actual profit or status of a project. Also, the "completed contract" method of accounting is entirely retrospective and provides no guidance for management. This is only one example of the types of allocations that are introduced to correspond to generally accepted accounting practices, yet may not further the cause of good project management.

Another common example is the use of equipment depreciation schedules to allocate equipment purchase costs. Allocations of costs or revenues to particular periods within a project may cause severe changes in particular indicators, but have no real meaning for good management or profit over the entire course of a project. As Johnson and Kaplan argue:

Today's management accounting information, driven by the procedures and cycle of the organization's financial reporting system, is too late, too aggregated and too distorted to be relevant for managers' planning and control decisions....

Management accounting reports are of little help to operating managers as they attempt to reduce costs and improve productivity. Frequently, the reports decrease productivity because they require operating managers to spend time attempting to understand and explain reported variances that have little to do with the economic and technological reality of their operations...

The management accounting system also fails to provide accurate product costs. Cost are distributed to products by simplistic and arbitrary measures, usually direct labor based, that do not represent the demands made by each product on the firm's resources.

As a result, complementary procedures to those used in traditional financial accounting are required to accomplish effective project control, as described in the preceding and following sections. While financial statements provide consistent and essential information on the condition of an entire organization, they need considerable interpretation and supplementation to be useful for project management.

Calculating Net Profit

As an example of the calculation of net profit, suppose that a company began six jobs in a year, completing three jobs and having three jobs still underway at the end of the year. What would be the company's net profit under, first, the "percentage-of-completion" and, second, the "completed contract method" accounting conventions?

Table. Example of Financial Records of Projects

Net Profit on Completed Contracts (Amounts in thousands of dollars)			
Job 1			$1,436
Job 2			356
Job 3			- 738
Total Net Profit on Completed Jobs			$1,054
Status of Jobs Underway	**Job 4**	**Job 5**	**Job 6**
Original Contract Price	$4,200	$3,800	$5,630
Contract Changes (Change Orders, etc.)	400	600	- 300
Total Cost to Date	3,600	1,710	620
Payments Received or Due to Date	3,520	1,830	340
Estimated Cost to Complete	500	2,300	5,000

As shown in Table, a net profit of $1,054,000 was earned on the three completed jobs. Under the "completed contract" method, this total would be total profit. Under the percentage-of completion method, the year's expected profit on the projects underway would be added to this amount.

Current contract price = Original contract price + Contract Changes

= 4,200 + 400 + 4,600

Credit or debit to date = Total costs to date–Payments received or due to date

	= 3,600–3,520
	=–80
Contract value of uncompleted work	= Current contract price–Payments received or due
	= 4,600–3,520
	= 1,080
Credit or debit to come	= Contract value of uncompleted work-Estimated Cost to Complete
	= 1,080–500
	= 580
Estimated final gross profit	= Credit or debit to date + Credit or debit to come
	= – 80. + 580.
	= 500
Estimated total project costs	= Contract price–Gross profit
	= 4,600–500
	= 4,100
Estimated Profit to date	= Estimated final gross profit × Proportion of work complete
	= 500. (3600/4100))
	= 439

Similar calculations for the other jobs underway indicate estimated profits to date of $166,000 for Job 5 and -$32,000 for Job. As a result, the net profit using the "percentage-of-completion" method would be $1,627,000 for the year.

CONTROL OF PROJECT CASH FLOWS

Project managers also are involved with assessment of the overall status of the project, including the status of activities, financing, payments and receipts. These various items comprise the project and financing cash flows described in earlier chapters. These components include costs incurred (as described above), billings and receipts for billings to owners (for contractors), payable amounts to suppliers and contractors, financing plan cash flows (for bonds or other financial instruments), etc.

As an example of cash flow control, consider the report shown in Table. In this case, costs are not divided into functional categories as in Table, such as labor, material, or equipment. Table represents a summary of the project status as viewed from *different components of the accounting system.* Thus, the aggregation of different kinds of cost exposure or cost commitment shown in Table 12-0 has not been performed.

The elements in Table include:

- *Costs*: This is a summary of charges as reflected by the job cost accounts, including expenditures and estimated costs. This row provides an aggregate summary of the detailed activity cost information described in the previous section. For this example, the total costs as of July 2 (7/02) were $ 8,754,516, and the original cost estimate was $65,863,092, so the approximate percentage complete was 8,754,516/ 65,863,092 or 13.292%. However, the project manager now projects a cost of $66,545,263 for the project, representing an increase of $682,171 over the original estimate. This new estimate would reflect the actual percentage of work completed as well as other effects such as changes in unit prices for labor or materials. Needless to say, this increase in expected costs is not a welcome change to the project manager.
- *Billings*: This row summarizes the state of cash flows with respect to the owner of the facility; this row would not be included for reports to owners. The contract amount was $67,511,602, and a total of $9,276,621 or 13.741% of the contract has been billed. The amount of allowable billing is specified under the terms of the contract between an owner and an engineering, architect, or constructor. In this case, total billings have exceeded the estimated project completion proportion. The final column includes the currently projected net earnings of $966,339.

- *Payables*: The Payables row summarizes the amount owed by the contractor to material suppliers, labor or sub-contractors. At the time of this report, $6,719,103 had been paid to subcontractors, material suppliers, and others. Invoices of $1,300,089 have accumulated but have not yet been paid. A retention of $391,671 has been imposed on subcontractors, and $343,653 in direct labor expenses have been occurred. The total of payables is equal to the total project expenses shown in the first row of costs.
- *Receivables*: This row summarizes the cash flow of receipts from the owner. Note that the actual receipts from the owner may differ from the amounts billed due to delayed payments or retainage on the part of the owner. The net-billed equals the gross billed less retention by the owner. In this case, gross billed is $9,276,621 (as shown in the billings row), the net billed is $8,761,673 and the retention is $514,948. Unfortunately, only $7,209,344 has been received from the owner, so the open receivable amount is a (substantial!) $2,067,277 due from the owner.
- *Cash Position*: This row summarizes the cash position of the project as if all expenses and receipts for the project were combined in a single account. The actual expenditures have been $7,062,756 (calculated as the total costs of $8,754,516 less subcontractor retentions of $391,671 and unpaid bills of $1,300,089) and $ 7,209,344 has been received from the owner. As a result, a net cash balance of $146,588 exists which can be used in an interest earning bank account or to finance deficits on other projects.

Each of the rows shown in Table would be derived from different sets of financial accounts. Additional reports could be prepared on the financing cash flows for bonds or interest charges in an overdraft account.

Table. An Example of a Cash Flow Status Report

Costs 7/02	Charges 8,754,516	Estimated 65,863,092	% Complete 13.292	Projected 66,545,263	Change 682,171
Billings 7/01	Contract 67,511,602	Gross Bill 9,276,621	% Billed 13.741	Profit 966,339	
Payables 7/01	Paid 6,719,103	Open 1,300,089	Retention 391,671	Labor 343,653	**Total** 8,754,516
Receivable 7/02	Net Bill 8,761,673	Received 7,209,344	Retention 514,948	Open 2,067,277	
Cash Position	Paid 7,062,756	Received 7,209,344	Position 146,588		

The overall status of the project requires synthesizing the different pieces of information summarized in Table. Each of the different accounting systems contributing to this table provides a different view of the status of the project. In this example, the budget information indicates that costs are higher than expected, which could be troubling. However, a profit is still expected for the project. A substantial amount of money is due from the owner, and this could turn out to be a problem if the owner continues to lag in payment. Finally, the positive cash position for the project is highly desirable since financing charges can be avoided.

The job status reports illustrated in this and the previous sections provide a primary tool for project cost control. Different reports with varying amounts of detail and item reports would be prepared for different individuals involved in a project. Reports to upper management would be summaries, reports to particular staff individuals would emphasize their responsibilities (eg. purchasing, payroll, etc.), and detailed reports would be provided to the individual project managers.

Of course, these schedule and cost reports would have to be tempered by the actual accomplishments and problems occurring in the field. For example, if work already completed is of sub-standard quality, these reports would not reveal such a problem. Even though the reports indicated a project on time and on budget, the possibility of re-work or inadequate facility performance due to quality problems would quickly reverse that rosy situation.

SCHEDULE CONTROL

In addition to cost control, project managers must also give considerable attention to monitoring schedules. Construction typically involves a deadline for work completion, so contractual agreements will force attention to schedules. More generally, delays in construction represent additional costs due to late facility occupancy or other factors. Just as costs incurred are compared to budgeted costs, actual activity durations may be compared to expected durations. In this process, forecasting the time to complete particular activities may be required.

The methods used for forecasting completion times of activities are directly analogous to those used for cost forecasting.

For example, a typical estimating formula might be:

$$D_f = Wh_t$$

where D_f is the forecast duration, W is the amount of work, and h_t is the observed productivity to time t. As with cost control, it is important to devise efficient and cost effective methods for gathering information on actual project accomplishments. Once estimates of work complete and time expended on particular activities is available, deviations from the original duration estimate can be estimated..

For example, Figure 9.2 shows the originally scheduled project progress versus the actual progress on a project. This figure 9.2 is constructed by summing up the percentage of each activity which is complete at different points in time; this summation can be weighted by the magnitude of effort associated with each activity. In Figure 9.2, the project was ahead of the original schedule for a period including point A, but is now late at point B by an amount equal to the horizontal distance between the planned progress and the actual progress observed to date.

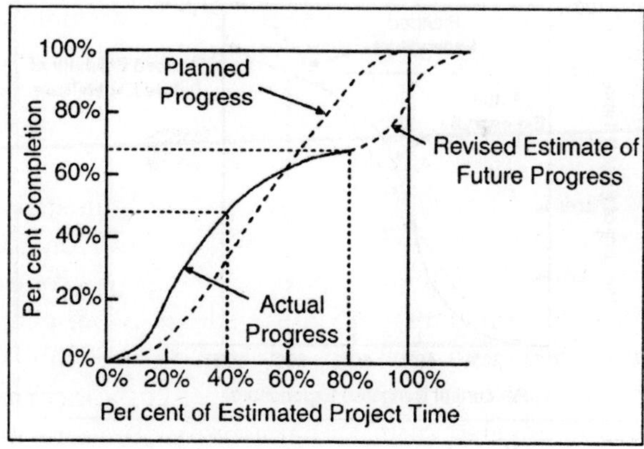

Fig. 9.2 llustration of Planned Versus Actual Progress over Time on a Project

Schedule adherence and the current status of a project can also be represented on geometric models of a facility. For example, an animation of the construction sequence can be shown on a computer screen, with different colours or other coding scheme indicating the type of activity underway on each component of the facility. Deviations from the planned schedule can also be portrayed by colour coding. The result is a mechanism to both indicate work in progress and schedule adherence specific to individual components in the facility.

In evaluating schedule progress, it is important to bear in mind that some activities possess float or scheduling leeway, whereas delays in activities on the critical path will cause project delays. In particular, the delay in planned progress at time t may be soaked up in activities' float (thereby causing no overall delay in the project completion) or may cause a project delay. As a result of this ambiguity, it is preferable to update the project schedule to devise an accurate protrayal of the schedule adherence. After applying a scheduling algorithm, a new project schedule can be obtained. For cash flow planning purposes, a graph or report similar to that shown in Figure 9.3 can be constructed to compare actual expenditures to planned expenditures at any time.

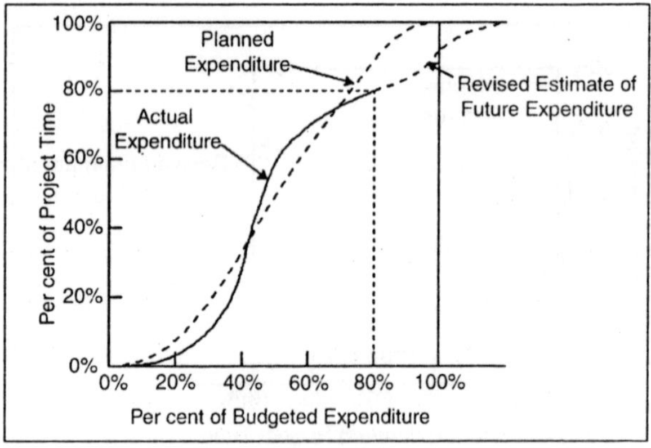

Fig. 9.3 llustration of Planned Versus Actual Expenditures on a Project

SCHEDULE AND BUDGET UPDATES

Scheduling and project planning is an activity that continues throughout the lifetime of a project. As changes or discrepancies between the plan and the realization occur, the project schedule and cost estimates should be modified and new schedules devised. Too often, the schedule is devised once by a planner in the central office, and then revisions or modifications are done incompletely or only sporadically. The result is the lack of effective project monitoring and the possibility of eventual chaos on the project site.

On "fast track" projects, initial construction activities are begun even before the facility design is finalized. In this case, special attention must be placed on the coordinated scheduling of design and construction activities. Even in projects for which the design is finalized before construction begins, *change orders* representing changes in the "final" design are often issued to incorporate changes desired by the owner.

Periodic updating of future activity durations and budgets is especially important to avoid excessive optimism in projects experiencing problems. If one type of activity experiences delays on a project, then related activities are also likely to be delayed unless managerial changes are made. Construction projects normally involve numerous activities which are closely related due to the use of similar materials, equipment, workers or site characteristics. Expected cost changes should also be propagated thoughout a project plan. In essence, duration and cost estimates for future activities should be revised in light of the actual experience on the job. Without this updating, project schedules slip more and more as time progresses. To perform this type of updating, project managers need access to original estimates and estimating assumptions.

Unfortunately, most project cost control and scheduling systems do not provide many aids for such updating. What is required is a means of identifying discrepancies, diagnosing the cause, forecasting the effect, and propagating this effect to all related activities. While these steps can be undertaken manually, computers aids to support interactive updating or even automatic updating would be helpful.

Beyond the direct updating of activity durations and cost estimates, project managers should have mechanisms available for evaluating any type of schedule change. Updating activity duration estimations, changing scheduled start times, modifying the estimates of resources required for each activity, and even changing the project network logic (by inserting new activities or other changes) should all be easily accomplished. In effect, scheduling aids should be directly available to project managers. Fortunately, local computers are commonly available on site for this purpose.

SCHEDULE UPDATES IN A SMALL PROJECT

A few problems or changes that might be encountered include the following:

- An underground waterline that was previously unknown was ruptured during the fifth day of the project. An extra day was required to replace the ruptured section, and another day will be required for clean-up. What is the impact on the project duration?
  - To analyse this change with the critical path scheduling procedure, the manager has the options of (1) changing the expected duration of activity C, General Excavation, to the new expected duration of 10 days or (2) splitting activity C into two tasks (corresponding to the work done prior to the waterline break and that to be done after) and adding a new activity representing repair and clean-up from the waterline break. The second approach has the advantage that any delays to other activities (such as activities D and E) could also be indicated by precedence constraints.
  - Assuming that no other activities are affected, the manager decides to increase the expected duration of activity C to 10 days. Since activity C is on the critical path, the project duration also increases by 2 days. Applying the critical path scheduling procedure would confirm this change and also give a new set of earliest and latest starting times for the various activities.
- After 8 days on the project, the owner asks that a new drain be installed in addition to the sewer line

scheduled for activity G. The project manager determines that a new activity could be added to install the drain in parallel with Activity G and requiring 2 days. What is the effect on the schedule?

- Inserting a new activity in the project network between nodes 3 and 4 violates the activity-on-branch convention that only one activity can be defined between any two nodes. Hence, a new node and a dummy activity must be inserted in addition to the drain installation activity. As a result, the nodes must be re-numbered and the critical path schedule developed again. Performing these operations reveals that no change in the project duration would occur and the new activity has a total float of 1 day.
- To avoid the labor associated with modifying the network and re-numbering nodes, suppose that the project manager simply re-defined activity G as installation of sewer and drain lines requiring 4 days. In this case, activity G would appear on the critical path and the project duration would increase. Adding an additional crew so that the two installations could proceed in parallel might reduce the duration of activity G back to 2 days and thereby avoid the increase in the project duration.

• At day 12 of the project, the excavated trenches collapse during Activity E. An additional 5 days will be required for this activity. What is the effect on the project schedule? What changes should be made to insure meeting the completion deadline?

- Activity E has a total float of only 1 day. With the change in this activity's duration, it will lie on the critical path and the project duration will increase.
- Analysis of possible time savings in subsequent activities is now required, using the procedures described in Section 10.9.

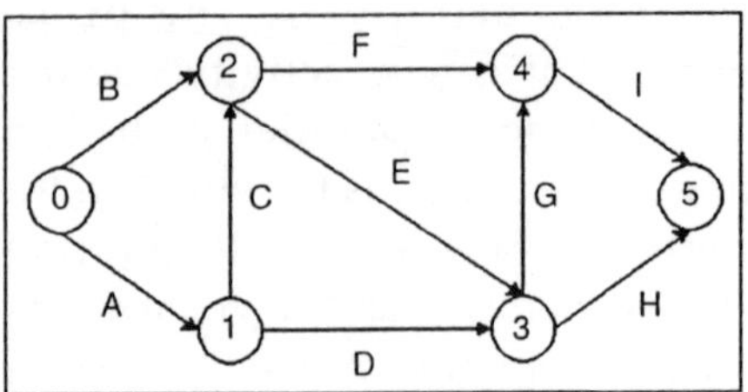

Fig. 9.4 A Nine Activity Example Project

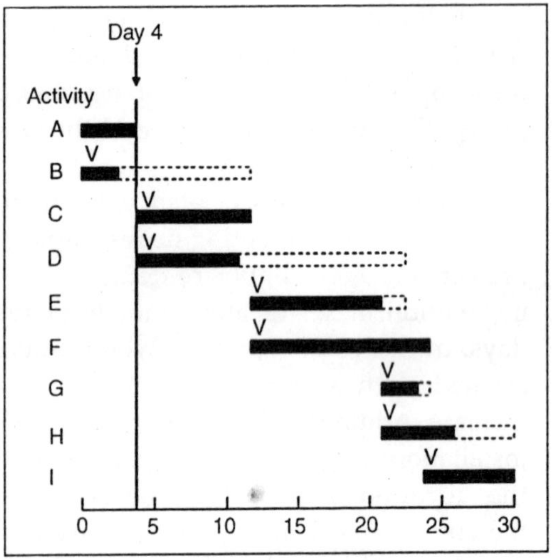

Fig. 9.5 Current Schedule for an Example Project Presented as a Bar Chart

As can be imagined, it is not at all uncommon to encounter changes during the course of a project that require modification of durations, changes in the network logic of precedence relationships, or additions and deletions of activities. Consequently, the scheduling process should be readily available as the project is underway.

RELATING COST AND SCHEDULE INFORMATION

The previous sections focused upon the identification of

the budgetary and schedule status of projects. Actual projects involve a complex inter-relationship between time and cost. As projects proceed, delays influence costs and budgetary problems may in turn require adjustments to activity schedules. Unanticipated events might result in increases in both time and cost to complete an activity. For example, excavation problems may easily lead to much lower than anticipated productivity on activities requiring digging.

While project managers implicitly recognize the inter-play between time and cost on projects, it is rare to find effective project control systems which include both elements. Usually, project costs and schedules are recorded and reported by separate application programmes. Project managers must then perform the tedious task of relating the two sets of information.

The difficulty of integrating schedule and cost information stems primarily from the level of detail required for effective integration. Usually, a single project activity will involve numerous cost account categories. For example, an activity for the preparation of a foundation would involve laborers, cement workers, concrete forms, concrete, reinforcement, transportation of materials and other resources. Even a more disaggregated activity definition such as erection of foundation forms would involve numerous resources such as forms, nails, carpenters, laborers, and material transportation.

Again, different cost accounts would normally be used to record these various resources. Similarly, numerous activities might involve expenses associated with particular cost accounts. For example, a particular material such as standard piping might be used in numerous different schedule activities. To integrate cost and schedule information, the disaggregated charges for specific activities and specific cost accounts must be the basis of analysis.

A straightforward means of relating time and cost information is to define individual *work elements* representing the resources in a particular cost category associated with a particular project activity. Work elements would represent an element in a two-dimensional matrix of activities and cost accounts as illustrated in Figure 9.6. A numbering or identifying system for work elements would include both the relevant cost account and the associated activity.

In some cases, it might also be desirable to identify work elements by the responsible organization or individual. In this case, a three dimensional representation of work elements is required, with the third dimension corresponding to responsible individuals.

With this organization of information, a number of management reports or views could be generated. In particular, the costs associated with specific activities could be obtained as the sum of the work elements appearing in any row in Figure 9.6. These costs could be used to evaluate alternate technologies to accomplish particular activities or to derive the expected project cash flow over time as the schedule changes. From a management perspective, problems developing from particular activities could be rapidly identified since costs would be accumulated at such a disaggregated level. As a result, project control becomes at once more precise and detailed.

Project Activity Group	**Cost Amount for Superstructure**					
	204.1	**204.2**	**204.3**	**204.4**	**204.5**	**204.6**
First Floor	X	X		X		X
Second Floor		X		X		X
Third Floor		X	X	X		X
Fourth Floor		X	X			X
Fifth Floor		X	X		X	X

Fig. 9.6 Illustration of a Cost Account and Project Activity Matrix

Unfortunately, the development and maintenance of a work element database can represent a large data collection and organization effort. As noted earlier, four hundred separate cost accounts and four hundred activities would not be unusual for a construction project. The result would be up to 400×400 = 160,000 separate work elements. Of course, not all activities involve each cost account. However, even a density of two per cent (so that each activity would have eight cost accounts and each account would have eight associated activities on the average) would involve nearly thirteen thousand work elements. Initially preparing this database represents a considerable burden, but it is also the case that project bookkeepers must record project events within each of these various work elements. Implementations of the "work element" project control systems have typically foundered on the burden of data collection, storage and book-keeping.

Until data collection is better automated, the use of work elements to control activities in large projects is likely to be difficult to implement. However, certain segments of project activities can profit tremendously from this type of organization. In particular, material requirements can be tracked in this fashion. Materials involve only a subset of all cost accounts and project activities, so the burden of data collection and control is much smaller than for an entire system. Moreover, the benefits from integration of schedule and cost information are particularly noticeable in materials control since delivery schedules are directly affected and bulk order discounts might be identified. Consequently, materials control systems can reasonably encompass a "work element" accounting system.

In the absence of a work element accounting system, costs associated with particular activities are usually estimated by summing expenses in all cost accounts directly related to an activity plus a proportion of expenses in cost accounts used jointly by two or more activities. The basis of cost allocation would typically be the level of effort or resource required by the different activities. For example, costs associated with supervision might be allocated to different concreting activities on the basis of the amount of work (measured in cubic yards of concrete) in the different activities. With these allocations, cost estimates for particular work activities can be obtained.

Index

D

E

F

G

H

I

J

U

V